The Arrogant Years

· The ·
Arrogant Years

ONE GIRL'S SEARCH FOR HER LOST YOUTH,
FROM CAIRO TO BROOKLYN

Lucette Lagnado

An Imprint of HarperCollinsPublishers

Grateful acknowledgment is made for permission to reproduce the photographs on the following pages: Page 9: courtesy of the Garzon family. Pages 23, 29, and 54: courtesy of Maria Cattaui, Cattaui Family Archives. Page 31: courtesy of His Royal Majesty, Ahmed Fouad II, from his family archives, Switzerland. Page 32: courtesy of Nimet Cattaui and Michel Alexane. Page 57: courtesy of Lucette Lagnado, Lagnado Family Archives. Pages 72, 79, 102, 104, 161, and 210: courtesy of César Lagnado, Lagnado Family Archives. Page 93: courtesy of Isaac Lagnado. Pages 94 and 96: courtesy of Rachel and Pico Hakim. Pages 142, 153, 204, 237, and 250: courtesy of Lagnado Family Archives. Pages 206, 207, and 228: courtesy of Ed and Maureen Kozdrajski. Page 212: courtesy of Betsy Raze. Page 284: courtesy of Laurie and Eli Bryk. Page 286: courtesy of Laurie Wolf Bryk and family. Pages 308 and 332: courtesy of César and Monica Lagnado, Lagnado Family Archives. Page 323: courtesy of Lagnado-Feiden Family Archives. Page 375: courtesy of Abdel-Hamed Osman.

HarperCollins books may be purchased for educational, business, or sales promotional use. For information please write: Special Markets Department, HarperCollins Publishers, 10 East 53rd Street, New York, NY 10022.

FIRST EDITION

Designed by Jennifer Daddio / Bookmark Design & Media Inc.

Library of Congress Cataloging-in-Publication Data has been applied for.

ISBN 978-0-06-180367-3

11 12 13 14 15   OV/RRD   10 9 8 7 6 5 4 3 2 1

To Douglas Feiden —

and in memory of

Leon and Edith, always Edith

She put on the first ankle-length day dress
she had owned in many years and crossed herself
reverently with Chanel Sixteen. . . . How good . . .
to be worshipped again, to pretend to have
a mystery. She had lost two of the great
arrogant years in the life of a pretty girl—
now she felt like making up for them.

—F. Scott Fitzgerald, *Tender Is the Night*

You are always talking about rebuilding the hearth.
What hearth? Tell me . . .

—Alphonse Daudet, *Le Petit Chose*

Blessed is the Lord who revives the dead.

—Hebrew daily prayer

• CONTENTS •

Prologue: The Avenger of Sixty-Sixth Street   1

Book One: The Curse of Alexandra

1. The Secret of the Pasha's Wife   21
2. The Alley of the Pretty One   35
3. The Bride Who Set Herself on Fire   59
4. The Bonesetter of Mouski   71
5. The Porcelain Dolls of Malaka Nazli Street   89

Book Two: Rebuilding the Hearth

6. The Legend of Agent Extraordinary   109
7. Passion Play on Sixty-Seventh Street   121
8. The Healing Powers of Iodine   135
9. The Errant Sister   151
10. The Passion of the Fast   157

Contents

11. The Messiah Is a Woman   167

12. The Tragedy of the Navy Blue Blazer   179

13. The Advance of the Little Porcelain Dolls   197

Book Three: Cities of Refuge

14. The Shrine on the Mountain   233

15. The Fall of Afterward   245

16. The Spring of Nevermore   265

17. The Princess of West 116th Street   277

Book Four: The Lady in the Pink Bow

18. The Lost Art of Penmanship   301

19. The Woman Against the Wall   327

20. An Earthquake in Cairo   341

Book Five: The Book of Lamentations

21. The Verse of Consolation   353

Epilogue: Inside the Pasha's Library   373

Acknowledgments   385

Bibliography   401

The Arrogant Years

The Avenger of Sixty-Sixth Street

ome September, the women's section of the Shield of Young David synagogue was once again crowded as families returned from their summer holidays, and those of us who had been left behind week after week to attend virtually empty services felt a bit like soldiers who managed to prevail on a hopelessly abandoned battlefield. Back from the family bungalow in the Catskills, Mrs. Ruben, the rabbi's wife, once again took her seat in the front, her flock of daughters in tow; the Ruben girls, each named after a biblical heroine—Miriam, Deborah, Rebecca, Rochelle—occupied an entire row behind the tall wooden divider that separated us from the men.

My mother, Edith, and I were among the first to arrive on Saturday mornings, and while we could have grabbed the front-row seats of the women's section for ourselves, more often than not we made our way to the second row, in deference to "la femme du Rabbin" as Mom liked to refer to Mrs. Ruben.

She never once called her by her name.

Rabbi Ruben's wife was a thin, stern figure who never raised her voice or lost her temper, yet still terrified me. She was vastly different from her husband, who was jovial and charismatic and whose passionate speeches from the pulpit held us in thrall.

Shortly after my family had left Egypt and moved to America two years earlier, the murder of Kitty Genovese dominated Harry Ruben's sermons. He'd constantly decried her brutal slaying, pounding the lectern as he recalled that dark March morning in 1964 when she'd screamed "Help me" but no one came to her rescue. He terrified us with his vision of a country where some forty neighbors could listen to a young woman's cries, shut their windows, and not bother to call the police.

Yet even at his most fiery and intense, Rabbi Ruben radiated goodwill and bonhomie. His wife, on the other hand, while perfectly proper and polite, seemed somehow disapproving even when she greeted us and wished us Good Sabbath. I had the feeling she didn't much care for any of us in this little immigrant congregation where fate had landed her. The Ruben daughters didn't budge from their chairs, and they tended to play among themselves. David, the lone son, occasionally wandered over because little boys were permitted to enter the women's section, and many darted in and out during services to confer with their mothers.

Our section was situated at the rear of the synagogue, though a portion of it jutted out into the sanctuary. It was small and rectangular shaped, surrounded on two sides—the front and the left—by a decorative beige wooden fence; the right side's natural boundary was the synagogue wall. Although it was an enclosed space, much like a pen, it felt more cozy than claustrophobic. Mom and I always took seats next to the divider on the left, where we could look out into the sanctuary and follow every prayer, every movement by the men.

I went to services dutifully every Saturday morning. I'd walk hand in hand with my mother from our house on Sixty-Sixth Street to the beige brick synagogue around the corner on Sixty-Seventh Street. Together, we would climb the dimly lit staircase to the sanctuary on the second floor, which was bathed in light from the windows as well as several crystal chandeliers.

Most of the women behind the divider were refugees from the Levant like us who had retreated to this small comfort zone whose name alone promised a buffer from the vagaries and pressures of the outside world: the Shield of Young David.

The wooden divider was an elegant if flimsy affair. After opening my prayer book for cover, I'd peer through its filigreed holes in the shape of diamonds and clovers and curlicues. I found it all so thrilling, the goings-on in the men's section. The men were constantly being called up to the elevated reading table at the center of the shul, so that there was a sense of continuous motion—they'd be marching, singing, bowing, putting on their prayer shawls, taking off their prayer shawls—whereas we mostly stayed put in our seats.

The men were the ones who led the services and chanted out loud. During the silent devotions, how I envied the way they prayed—with such focus and single-mindedness—their heads and shoulders wrapped in their soft white shawls. Surely, they enjoyed a special relationship with God.

I was anxious to trade places with them, to be the one to lead prayers and lift Torah scrolls high in the air. I noticed that whenever a man was called up to the Holy Ark at the front of the synagogue, his wife and daughters jumped to their feet and paid tribute to him by remaining standing until he returned to his seat. Would anyone ever stand up for me?

In my mind, there were two worlds—the gossipy, trivial, inconsequential world of the women's section and the solemn, purposeful world beyond it, the world where men sat in vast and airy quarters communing with God. The world that I longed to join and where I felt I belonged. The world beyond the divider.

Our section became more cramped with every passing week. As we approached the High Holidays, there was barely an empty seat.

In anticipation of the holidays, all the women began to dress up; and everyone, from dowdy matrons to toddlers, paraded in their finest clothes. Our little house of worship turned into a fashion runway. Everyone eyed what the others were wearing.

It was a challenge I couldn't resist.

In the fall of 1966 when I turned ten, my heroine was Emma Peel, the British secret agent who managed to be stylish and lethal at the same time, delivering karate chops and judo flips while clad in skintight black leather. Mrs. Peel was the star of *The Avengers,* a show that I watched obsessively each week on our new TV set, our first major purchase since coming to America.

I liked to imagine myself as a world-class spy, an international woman of intrigue. Consumed with Emma Peel, I tried to emulate her speech and manner and adopted her wry half smile and her hairdo. I longed for her strength, her courage, her wit, her intellect, and above all, her wardrobe.

Alas, I couldn't find Mrs. Peel's sleek leather jumpsuits—the epitome of the new London chic—on Brooklyn's Eighteenth Avenue, the strip of bargain stores where we usually shopped, or at the few department stores where we dared venture like S. Klein's on Union Square, Mays on Fulton Street, or Gimbel's Basement. None of the wonderful clothes she wore, not the jaunty hats or the hip-hugger slacks or the geometric mod dresses, seemed available to me in my universe of discount outlets, though I searched and searched and wondered when I would attain the Avenger's elegance and flair.

At last, my runaway older sister came to the rescue.

Shortly after my birthday, I arrived at the women's section dressed in a dashing green woolen blazer with gold buttons and a gold crest with a matching forest green velvet Carnaby hat, both gifts of Suzette who had left home two years earlier, shortly after we'd settled in New York. She compensated for her departure by lavishing me with expensive clothes. I wore the hat, which resembled a newsboy's cap, tilted at an angle and let my hair grow long, resisting my mom's entreaties to trim it.

I wanted it the precise length as Emma Peel's—slightly past my shoulders in a soft flip and swept off my forehead.

I'd make a grand entrance on Saturday mornings in my blazer and carefully angled hat and take my seat by Mom, feeling confident and self-assured and in my view decidedly more elegant than any other inhabitant of the women's section. As more women arrived over the course of the morning, I'd leave my mother's side and stride up and down our little enclave to be embraced by some, patted on the head by others, and—I hoped—admired by all.

What do you want to be when you grow up? I'd be asked.

An Avenger, I always replied, without missing a beat.

The show's introduction became my mantra. I would recite it out loud to anyone who would listen, pleased with myself for having memorized it. I attempted a British accent that must have sounded jarring, coming on top of the French accent I couldn't quite shed. *"Extraordinary crimes, against the people and the state, have to be avenged by agents extraordinary. Two such people are John Steed, top professional, and his partner, Emma Peel, talented amateur, otherwise known as The Avengers . . ."*

Of course, it wasn't clear what there was to avenge in this cosseted little world of mine. There were no extraordinary crimes here.

There weren't even many ordinary crimes. Sixty-Sixth Street was remarkably safe—the kind of street where children played day and night without a care, where physical danger seemed remote. Bensonhurst was a staid, working-class area whose residents, mainly newly arrived immigrants, felt so removed and disenfranchised from the rest of New York that everyone around me called Manhattan "The City."

My universe consisted of about a dozen blocks, bounded on the west by Eighteenth Avenue, the lively discount shopping district my mom adored, and on the east by Bay Parkway, the vast and slightly more opulent boulevard of banks, luncheonettes, and stationery stores that my father favored. Wedged in between were my elementary school, my synagogue, my friends, and Key Food, my first American supermarket whose shelves I liked to scour and whose aisles and sawdust-covered floors I found enticing.

I much preferred the world of the Avengers—the heady world of 1960s London, where one woman emblemized all I wanted to be on this earth. I prayed for the day when I would be asked to step in and avenge some extraordinary crime.

Every Saturday as I walked with my mom to services, I considered the possibility that my skills as an Avenger would be urgently required. I imagined a hostage situation—arriving one morning and finding the rabbi and the other men being held at gunpoint by vicious marauders—mad scientists, master destroyers, military megalomaniacs. I pictured myself rushing to the front of the synagogue exactly like Emma Peel and, with some elegant judo moves, disposing of the bad guys who had dared invade our little world.

But we'd walk in to the same placid scene we found every week. There was Mr. Menachem conducting the prayers with his usual pleasant cadences while Rabbi Ruben sat, as always, impassive in his armchair on the makeshift stage at the front, and I took my seat next to Mom by the divider.

I was usually blissfully content in my chair. The universe as defined by the wooden partition was one of the few places where I could be myself, where I felt at ease, and where that sense of not belonging, of being different and foreign that had haunted me since leaving Egypt, vanished.

But I also chafed at the divider. The more I watched the men, the greater my longing to join them, sit next to them, worship at their side.

The extraordinary crime was there, in front of me, I decided. The extraordinary crime was the divider itself.

I knew what I had to do: I had to become the Avenger of the women's section. I wanted to demolish our wooden enclosure, to smash it into a thousand pieces, to strike it down with some deft karate chops the way I knew Emma Peel would if given the opportunity.

A visitor coming to our little congregation for the first time would be impressed by the strict separation between the sexes, the fact that men seemed totally consumed in the prayers even as women were kept at a safe distance. But if they spent a morning or two with us, they would realize that our little plywood fence did nothing to stop flirtations or prevent the passions that inevitably flared up despite the barrier.

And so, no, the men weren't nearly as caught up in their prayers as they seemed. I noticed how they always perked up when a pretty woman walked in; all male eyes would be on her, divider or no divider. The single men would make a beeline toward her when services were over. That is when they were permitted to wander into our section, which had a long table set up in the back for the kiddush meal. This often lavish luncheon was prepared each week by Gladys, our own in-house chef. Gladys prayed devoutly, cried a great deal, then snuck away to the kitchen halfway through services.

We knew that she was both arranging a feast and partaking of it.

Later at the table, a young man could approach his object of desire, strike up a conversation, maybe even boldly ask her to meet him later that night, once Sabbath was over, because life for us stood still until sundown; only then could we pick up where we'd left off.

I made a mental note of all these comings and goings like the secret agent I hoped to be. I would try to spot a budding romance or a nascent entente, and I liked to think I was the first to notice signs of a disintegrating relationship.

Every one of us in the women's section thought about romance—perhaps as much as we thought about God. There were the young single women hoping to find their soul mate among the sea of men beyond the divider, anxious that time was passing them by since any woman over twenty-five was considered unmarriageable in our community. There were the mothers with marriageable daughters, wondering if their girls would be able to find a match fast enough in the face of the unrelenting pressure. There were the teenagers who were on tenterhooks, since eighteen was the requisite age to be engaged and married.

Then there were my friends and I—too young to be in love and suffer from emotional entanglements, or so the adults around us thought.

My mother's closest friend was a recent Moroccan immigrant named Madame Marie, whose twelve-year-old daughter, Celia, was my friend. Celia was unruly—or "wild" as we liked to whisper—and Madame Marie was constantly negotiating between

Celia's tenth birthday; she is the one drinking from the cup; Moshe, her brother, is next to her, and Loulou is next to Moshe. Edith is next to Madame Marie, who is holding the baby, her nephew.

her headstrong daughter and her husband, a stern man who wore wire-rimmed glasses and was prone to getting angry. A gentle soul, she'd confide her woes to my mom in anguished whispers during services. My mother would comfort her while complaining that I, too, caused her sleepless nights.

"Je suis folle d'inquietude pour Loulou," Mom would tell her. "I am worried sick about Loulou."

It was America, they agreed, a country that could ruin young girls if they weren't careful. They sighed, wistfully recalling idyllic girlhoods in the Levant, where a daughter knew exactly what was expected of her and her parents also knew, and life made so much more sense than here in New York.

very gathering has its bullies, and our synagogue was no exception. My tormentor among the men was a teenage boy named Charlie who relished standing up every few minutes to approach the women's section, prayer book in hand. He and his friends, a group of teenage boys, took turns yelling at us with an authority no one had ever conferred on them. "Quiet, ladies, quiet—The men are trying to pray," they'd cry. There would be a hush and they'd walk away, smug at having cowed us into submission.

We also had our oppressors behind the divider, none as formidable as Mrs. Menachem. She was the cantor's wife, which gave her a certain degree of clout, much like the *rebbetzin*. But whereas Mrs. Ruben wielded her authority quietly, subtly, Sarah Menachem liked to let everyone know she was the boss.

My distress over the divider was hardly a secret in the women's section. Confident and exceedingly arrogant in my green blazer and matching hat, I loved to expound on my views to anyone who would listen, and that included proclaiming my contempt for the *mehitzah*, as the barrier was called in Hebrew.

Mrs. Menachem knew of my feelings—my resentment of the divider and other religious conventions I thought were unfair.

"You are a silly, silly little girl," she hissed at me one Saturday morning in front of all the women, "who is trying to change the world."

For Mrs. Menachem, to challenge the existing order was a desecration. She would never think of questioning any sacred ritual or tradition, let alone the need for a rickety separation, nor could she fathom why I kept railing about it. She feared my influence on the other children and became convinced I was leading them astray.

Mrs. Menachem believed that I needed to be cut down to size. I had to be put in my place now—immediately—before it was too late and I had harmed myself and the other little girls who looked up to me. She hated what I was trying to do to her peaceable little world.

Nineteen sixties America, with its emergent culture of rebellion and change for the sake of change, its angry youth and defiant women, had to be shut out at all cost, even if that meant erecting larger, taller, less porous, more impervious barriers to protect us from its dangers.

Mrs. Menachem was our watcher at the gate, standing guard at the entrance of the women's section, making sure that it would stay free of outside influence, ascertaining that the divider was still solid enough to keep out interlopers.

We were sworn enemies, Mrs. Menachem and I, engaged in a fight to the death. I regarded her as Emma Peel would a ruthless opponent.

I also had my protectors, none as vigilant as Gladys—sad, overweight, painfully sweet Gladys. Because she loved to eat, food was always plentiful and our luncheons and afternoon snacks were more lavish than those of other congregations. Saturday mornings, Gladys prepared an enormous bowl of lemony tuna fish salad, which formed the centerpiece of the light buffet lunch that followed the service or was served later in the afternoon. On holidays, she cooked more elaborately, typically, southern fried chicken. It was my first taste of American cooking, and I found it transporting. Biting into those coated drumsticks was as close as we'd come to assimilation in our little world that was trying so desperately to shut out America. Gladys embraced me, fed me extra drumsticks, treated me as if I were her daughter.

But why was she always crying in her prayer book? Why did she break down and become so distressed some mornings that no one could console her, not even her younger sister Fortuna who was devoted to her and sat next to her at all times? No one would tell me; it was one of the mysteries of life in the women's section

As I peered through the holes of the divider I was on the lookout for only one person. There he was in his elegant maroon blazer with gold buttons, absorbed in his prayers in a way I could only pretend to be. Thirteen-year-old Maurice had been the object of my affections since I'd first started coming to the Shield of Young David.

What could I do to make Maurice notice me? That was the question that consumed me. I began formulating a plan. If I put my mind to it, I was sure I could figure out a way that would free me and the other girls forever from the divider.

It was 1966, a good year to rebel and shatter barriers. London, Mrs. Peel's London, had invaded our own hopelessly sober culture. Skirts were shorter and flashier and more daring, not merely mini but micromini, made of vinyl that came in shiny yellow and fire engine red and black or spiffy white. British bands were all the rage. British shows led by my beloved Avengers dominated television.

I felt supremely self-confident; I kept repeating the mantra:

Extraordinary crimes against the people and the state ought to be avenged by agents extraordinary . . .

*W*hy can't we sit with the men?" I asked the other little girls who worshipped with me every Saturday. Halfway through the services, I left my mother and huddled with my friends in the back, where we enjoyed some privacy. I shared my views with them. "Why should we have to sit behind a barrier?"

Diana, my closest friend, nodded in agreement; she was a year younger than me, loyal and brave, and seemed to trust implicitly what I said. The Cohen sisters, Gracie and Rebecca, were a bit more skeptical but willing to be good sports. Celia, who was two years older than us, merely smiled. Celia chafed against her strict upbringing.

For her, this was an opportunity.

I had the outlines of a plan. If I couldn't break the divider with a single karate chop, I was going to render it meaningless and beside the point. I gathered my friends around me in a corner of the yard where no one could overhear us, feeling like a general briefing his troops on an intricate and highly covert maneuver. I knew that what we were about to do was fraught with peril, that we had to proceed carefully, methodically.

We were going to infiltrate the main sanctuary and sit with the men. I swore all my friends to secrecy. No one in our families— neither our siblings nor our parents nor any other adults—could know. This was the plot:

On a typical Saturday morning, we would begin by placing our chairs outside the entrance to the women's section. We would sit quietly and pray in that gray nether zone that was neither within the confines of the wooden divider nor inside the main sanctuary where the men congregated. Then, each week, we would quietly advance, pushing our chairs several inches, maybe a foot or two, until we were smack in the sanctuary and seated with the men.

The key was to proceed stealthily and so gradually that nobody would notice.

My thought was to confront the congregation with a fait accompli. One day the women would learn they had lost us, and the men would look up and realize that we were among them. But by then it would be too late: They would have been conditioned into accepting us in their midst.

Because the High Holidays were upon us, we were in a period when our mothers were so distracted they left us alone. They were caught up in their usual frenzy of cooking and baking while the synagogue itself was so crowded nobody took much notice of us. I was sure no one would pay attention as we began our maneuvers.

Even on Yom Kippur, when we were supposed to be atoning for our sins, the women behind the divider turned that solemn day into a gossip fest. While the men prayed and pounded their hearts with their fists in a gesture of repentance, our mothers and sisters talked and talked about subjects that weren't in the least bit holy. I heard so many rumors being spread that it would have taken another Yom Kippur to atone for the sins of that day alone.

We met during services in the courtyard and plotted and schemed. None of us seemed to consider the possibility we would fail, least of all me.

"We have to move little by little," I reminded my friends. "We have to do it in such a way that they won't notice." Once the holidays were over and the crowds had thinned out a bit, I started searching for the perfect window to execute my plan.

One Saturday morning I left my mother's side and carried a folding chair to the entrance of the women's section. It was still early, and there was only a sparse crowd of men and virtually no women. I placed the chair close to the entrance but technically outside the boundaries of the divider.

Then, I sat down, opened a prayer book, and began to read.

To my amazement, nobody seemed to notice or even care.

The men nodded as they made their way past me to their seats in the sanctuary. The women cheerfully waved hello as they filed in and went to sit, dutifully as always, in their pen surrounded by the wooden fence. Celia and her family arrived, and I noticed Moshe, Celia's little brother, watching us. It could have been any other Saturday.

My friends—Diana, Celia, and a couple of the Cohen sisters—followed my lead. They took their chairs outside the women's section, placed them next to mine, and opened their prayer books. Gracie had brought her younger sister, Rebecca. We didn't even dare look at one another. Instead, we stared at the pages of our prayer books and tried to focus on the words on the page.

The insurrection had begun.

We were scared to death, of course. We avoided making eye contact with anyone entering or leaving the synagogue, and for once, we didn't even whisper to each other. We simply tried to blend in, rising when the rest of the congregation rose, chanting when everyone chanted, desperately hoping not to attract attention.

One of us giggled and that made the rest of us want to giggle, too. Even so, the first maneuver went off without a hitch.

We managed to get through the service perched in this nether zone. We were outside the divider, no longer with the women, though not exactly with the men, either. Then, at the end of the prayers, we made our way back inside the women's section and over to the kiddush table, as if nothing had happened, and devoured forkfuls of tuna fish salad that Gladys had made that morning. It tasted especially delicious, and I ate with relish—elated at what we'd managed to pull off and sure we would prevail.

In a way we already had. We had breached the wooden barrier. We had left the women's section.

The following week, I got there a bit earlier, carried my chair outside the divider, and pushed it closer to the well of the main sanctuary. It was only by a couple of inches or so, not much more than the previous Saturday, but that was all part of my grand conspiracy. My friends joined me and positioned their seats near mine as we had agreed.

Once again, nobody seemed to mind, and we were left alone.

The women, even my own mother, seemed unaware of what we were doing. I couldn't help noticing, though, that as the men filed in, a couple of them frowned, surprised at seeing us in such an odd place, as if suspended in midair, not seated with the women but not with them, either.

But we were little girls, no doubt absorbed in some amusing new little girl game. What harm was there in letting us play?

After nearly a month of these forays, I felt confident enough to take more decisive action. I decided to position our chairs several feet forward so that we were almost—almost—inside the open well of the sanctuary, not far from the altar where Mr. Menachem read the prayers in his pleasant singsong voice.

It was a bold move. We were now in plain sight of the men. It was hard for us to keep from smiling and restrain our glee. We were so close to reaching our goal—my goal. I had dreamed of this moment. We were at last equal partners with the men, with no wooden partition to block us.

I was secretly astonished my scheme had worked as well as it did: Had we really conditioned the men to have us in their midst? There we were, a ragtag army of little girls seated on wooden folding chairs in the heart of the sanctuary defying all convention, having broken free of the enclosure reserved for us and our mothers.

We prayed silently and held our breath. When the Torah scrolls were brought out, we didn't try to touch them as they came around, though they were now within easy reach of our hands. We simply stood demurely and blew kisses at them with both our hands, even as our mothers inside the women's section grasped at them from the holes in the divider.

It was a historic moment as far as I was concerned. I hadn't smashed the divider as Mrs. Peel would have done, and yet maybe I had.

Mr. Menachem began to read the weekly portion. From his small stage, the rabbi sat in his thronelike chair quietly observing the room as he did every Saturday, as if nothing were amiss. The service seemed to flow at its usual indolent, otherworldly pace.

Then, seemingly out of nowhere, the counteroffensive began.

Charlie and the other teenage boys—my oppressors, whose self-imposed mission was to maintain order—were suddenly on to us. They'd realized our scheme. They approached us, menacingly waving their prayer books like a weapon. "Get back, get back into the women's section," they were shouting. Someone yelled, "Haram, haram," the Arabic word for sin.

A number of the other men began screaming at us, too. "Haram, haram"—"Sin, sin" they cried so that it felt as if we were being overrun by an angry mob. Everyone was yelling at us, and we found ourselves surrounded by a group of boys and men who kept ordering us to retreat, who kept trying to shoo us back into the women's section.

Mr. Menachem, who had made a valiant effort to keep reading throughout the brouhaha, suddenly stopped. Rabbi Ruben remained quiet, surveying the scene. He didn't weigh in; he didn't order the boys to stop nor did he declare that we could remain in the men's section. He didn't say a word.

The women were also silent, watching the strange scene unfold, too stunned to speak. They didn't try to stand up for us or order their sons and nephews and husbands to leave us alone. They simply sat and stared as we were forced to lift our chairs then and there and carry them back inside the women's section.

I didn't dare look at my mother. I stole a glance at Maurice, standing quietly in his usual corner, impassive as always, observing the scene. And I noticed Moshe, Celia's younger brother, also staring, a witness to our terrible debacle.

It was all over within minutes. Mr. Menachem, after making

sure all was quiet, resumed the weekly Torah reading. Charlie and the other teenage boys sat down again. And my friends and I took our old seats behind the divider. I felt crushed and mortified. From my chair, I could see Mrs. Menachem; she looked angrier than I'd ever seen her. But she stayed silent, glaring into her prayer book.

I resolved then and there to leave the Shield of Young David.

"I'm never coming back," I told my friends.

The following Saturday, I made my way to the Greek shul, a small synagogue on Sixty-Fifth Street that didn't have a divider, or not much of one. A group of mostly older men and women merely sat on different sides of a sanctuary that looked like an auditorium, with only an aisle and a white veil curtain to separate them. It was, in its own way, egalitarian, far more so than any congregation I had ever attended. I tried to follow the service, but I felt restless and alone.

I missed my friends. To my surprise, I missed the women's section; I even missed the wooden divider.

A couple of weeks later, I returned to the Shield of Young David and quietly took my seat next to Mom.

"Loulou, s'il te plaît reste tranquille," was all that she said; Loulou, please try to keep still.

My friends welcomed me back as if nothing had happened, and nobody ever spoke of our rebellion again.

BOOK ONE

The Curse

of

Alexandra

CAIRO: 1923–1963

1

The Secret of
the Pasha's Wife

Cairo was never as hopeful as at that moment when its leading feminist, Hoda Shaarawi, stepped off a train at the Ramses station on Malaka Nazli Street and tore off her veil in a gesture of defiance. The year was 1923, King Fouad was in power, and there was change in the air—this ancient city was rapidly modernizing and nowhere was that more apparent than in the women who were asserting their freedom and independence for the first time ever in a Muslim culture. Hoda's friends who came to greet her were stunned by her action, but then they, too, yanked the veils from their faces and cast them aside in solidarity and, voilà, a liberation movement was born among the least liberated women in the world.

A few years later, a woman lifting her veil in Cairo once again caused an enormous stir. This time, she was made of granite—a tall formidable statue called *Egypt's Awakening* that depicted a peasant girl removing the veil from her face even as her hand rested on the head of the Sphinx.

The message was clear: The land of the pharaohs was forging a brand-new destiny for itself.

That sense of energy and inexorable social change—of barriers being torn down and age-old traditions being upended—was felt throughout Cairo of the 1920s and 1930s, even in the popular music. The crooner and matinee idol Mohamed Abdel Wahab was attracting enormous audiences performing songs with a distinctly Western influence. In a shocking departure from traditional Middle Eastern music, Abdel Wahab included a piano and even a saxophone in his orchestra. While King Fouad was firmly in control, there was still open and vigorous political debate and an outspoken opposition party. As yet another sign of how liberal the culture had become, Jews and Muslims and Christians mixed and mingled without paying much heed to religious differences.

Jews, in particular, had never fared better in a society that in many ways emblemized tolerance. They were rising to the top and becoming not simply ministers but pashas and beys. In the pecking order of titles conferred by the king, there were effendis, a grand honor; beys, an even grander honor; and pashas, the grandest honor of all. Influential Jews were now involved in shaping every sector of society, from banking to agriculture, from commerce to education.

At Fouad's court, a woman—a Jewish woman, at that—now held more power than the queen herself and had emerged as a favorite of the Muslim king, one of his most faithful and trusted advisers.

Madame Alice Suarez Cattaui Pasha was officially *la grande dame d'honneur*—chief lady-in-waiting to the court. But everyone knew she was much more than that. Even while assisting Queen

Madame Alice Cattaui Pasha as a young woman in Egypt.

Nazli, Madame Cattaui had become the confidante of the king, so that she was in the unusual position of enjoying the ear of both of Egypt's monarchs.

King Fouad depended on this elegant older woman for her guidance and judgment. He let her decide which visiting dignitaries he or Nazli should receive on a particular day as well as those minor aristocrats who could be safely ignored. When there were dinners at the palace, she was in charge of the complex seating arrangements. Because she effectively controlled access to the king, deciding who sat near him and who didn't, the pasha's wife wielded unprecedented power in Egypt.

While the poor queen, who had a very testy relationship with Fouad, was said to be virtually a prisoner of the palace, her chief lady-in-waiting was attending glittering soirees all over Cairo. Madame Cattaui was seen around town at ballets and galas and premieres. Foreign diplomats and their wives knew to call on her and woo her because she was the gatekeeper to the throne and could help them wangle an invitation to the palace.

Fouad himself sent her effusive notes of gratitude in French. It was the language of the aristocracy, and truth be told, the ruler of Egypt, so European in his tastes and manners, and fluent in Italian as well as French, could barely speak a word of Arabic.

Of course, it wasn't her skills alone that had originally propelled her into this position. Alice Cattaui was born into the Suarez family, one of the wealthiest in Egypt. Her husband, the pasha, was an engineer by training who had made his own vast fortune running the country's lucrative sugar-refining concern, Kom Ombo. Yussef Cattaui Pasha was a founding board member of Banque Misr, the first Egyptian bank in a country where all the financial institutions were foreign owned. He was also president of the Jewish community, a mission he took to heart as did his wife, because ministering to Cairene Jews who were destitute was part of the Cattaui heritage, and essential to the family's sense of noblesse oblige.

The neo-Gothic "Villa Cattaui" wasn't the largest mansion in Garden City, but it was certainly the most exotic. In this dreamlike corner of Cairo favored by the British, it stood out for its turrets and vaulted arches and stained-glass windows, but what made it unique was a library that housed more than sixty thousand volumes, handpicked by their bibliophile owner. The library was the pasha's great love, and he had designed it to be the most sumptuous part of the villa. It had its own wing with rows and rows of intricately de-

signed wooden cases; behind glass were sets of leather-bound volumes Yussef Cattaui had acquired throughout his life—rare first
editions of any and all subjects that interested him.

The pasha and his wife both entertained frequently. Madame
Cattaui was a striking figure, small with impeccable posture. Her
clothes were bought in Paris (though she did, of course, have favored *couturieres* in Cairo), and she was rarely seen without her
multiple strands of pearls and the special Queen Nazli pin en-
crusted with emeralds and diamonds and rubies that she wore like
a badge of honor. The brooch signified she had unfettered access
to all the royal palaces.

At home at Villa Cattaui, cooks and nannies and governesses
and housekeepers were there to attend to every need. When the
Cattauis' two grown sons, Aslan and René, got married, they had
their wives move in with them. Each brother took over a floor of
the mansion. That was the way you lived in Cairo, whether you
were in the humblest or the most elegant part of town. It was a culture where families—affluent, poor, or that small percentage that
was middle class—stayed together. It wasn't unusual for multiple
generations to reside under one roof, though the roof was rarely as
luxurious as the one at 8 Ibrahim Pasha Street.

The pasha's wife was in perpetual motion. The court, the
galas, the state dinners, the pressures of attending to a difficult
and headstrong queen while fulfilling her duties to the king would
have been exhausting for most human beings, but Madame Cattaui seemed unstoppable as she raced across Cairo on one royal
mission or another.

Friday afternoon was set aside for high tea, when she received
important women passing through—the princesses and other members of European nobility who were visiting Egypt and craved an
audience with Queen Nazli.

Afterward, it was quiet at the Cattaui residence, as it was in different parts of the city. Cairo was so respectful of its Jewish population that even *la bourse*, the stock market, shut down in observance of the Sabbath, as did many banks.

Sunday night, there was a festive meal at home with the family and selected friends. Everyone dined on plates rimmed with gold, featuring the distinctive Cattaui monogram. Guests couldn't help noticing the grand piano in the main drawing room, which was covered with pictures of the European aristocrats who had met with the pasha's wife.

Madame Cattaui was completely at ease in Cairo's high society and indeed dominated it. But she was equally committed to her work in a very different part of Cairo, the older neighborhoods where the Jewish communal institutions were situated. She took a special interest in the schools the Cattaui family had founded and were still bankrolling; her work there was as important as her duties toward the king and the queen.

It was only a short car ride from Garden City to the heart of Daher and Abbassiyah, yet it was a journey few residents of the leafy villas made, at least not regularly, and that is why the pasha's wife stood out.

She was a constant visitor to L'École Cattaui, the little Jewish private school in Sakakini the family had founded, and which prided itself on giving the finest and most rigorous education in all of Cairo. She'd also go regularly—every Tuesday in fact—to the Sebil, the massive communal Jewish school that catered to children who lacked means.

The students at the Sebil lived for Tuesdays and the glimpse they caught of the striking woman in silk and pearls. While she was clearly a grande dame, she had a gentle air about her. If children looked thin, or came to school in threadbare clothing, she would go over to them and then gently quiz their teachers. Were they in need? Could she possibly help?

Le Sebil guaranteed a free lunch to all its students. There were many stories of children going hungry, whose parents couldn't afford to give them so much as a sandwich to take to class. The school made sure they ate, though it was deeply humble fare. Come noon, in the large dining hall, hundreds of pupils sat at long tables with wooden benches, and the kitchen staff would arrive carrying big steaming pots with the day's offerings—a bowl of string beans with rice, or stewed potatoes, perhaps some lentil soup.

That was the menu day after day—except Tuesday. On a typical Tuesday morning, there'd already be a buzz about lunch. If the children heard that meat was on the menu, they'd loudly exclaim, "Madame Cattaui is coming."

The pasha's wife would arrive, usually with one other society woman in tow. The two would stand in the large cafeteria, carefully inspecting what was served. The school took extra pains to prepare a special meal those days the VIP visitors were expected.

They were the ladies of Tuesday, and they made it a point never to sit down. The children would see them walking up and down the long tables making sure there was enough to eat and that the dishes were clean. If a child wasn't eating, Madame Cattaui would coax him to finish his meal.

Once lunch was over, the children would sing the traditional after-meal hymn. But there were times a child had lost a parent or close relative. They were encouraged to recite the Kaddish, the Hebrew prayer for the dead. The children noticed the pasha's wife listening intently as they prayed, and it was as if she were praying with them.

She'd return days later and head for the courtyard. Needy children were pulled aside by their teachers—the students who wore torn shoes and dirty clothes and whose families lived in the ancient Haret-el-Yahood, the Jewish ghetto.

Madame Cattaui proceeded to give out *des sandalettes*—small,

inexpensive leather sandals. She also distributed packets of clothing, usually the aprons children were required to wear as a uniform—along with notebooks and pens and any other necessities
their parents couldn't afford to buy them.

There was a solemn, ceremonial quality to the affair; it was
supposed to be discreet, but everyone at school could see what was
going on in the courtyard and follow how she handed out the allot-
ment of *sandalettes,* and which children were lucky—or unlucky—
enough to receive them.

*W*hen she wasn't on official duty, Alice Cattaui was a different person. The opulent clothes and couture hats and
jewels came off and were replaced by simple black dresses; that is
what she preferred to wear at home alone, or to the market to buy
vegetables and groceries.

Now that was a task that could have been handled by the servants, yet the pasha's wife insisted on doing it herself.

She raised her children—then her grandchildren—with a message of stoicism, self-discipline, and tough love.

"Never give in to despair," she told her granddaughter Nimet
again and again.

To her family, her behavior could be mystifying. She looked
like a woman in mourning, even on festive occasions. And that is
exactly what she was—a woman observing the death of a loved
one, except in this case the loved one had died decades earlier.

The rule at Villa Cattaui was never to talk about Indji, the only
daughter of the pasha and his wife. She was rarely mentioned by
name, yet she seemed to be everywhere, lurking in every corner
of 8 Ibrahim Pasha Street, so that the lives of all of its inhabit-
ants were affected by her, even the grandchildren who had never
known her.

Born in 1888, the oldest of the three Cattaui children, Indji was doted on from the start. The photographs and portraits around the house and in the albums attested to her privileged status—dozens and dozens of images of a beautiful child taken from the time she was an infant, and always, or almost always, dressed in white.

As she grew up, the outfits became ever more intricate and luxuriant. One year, Indji posed in an Oriental costume, standing next to a Chinese vase. A year later, she was pictured in a knee-length white dress, high-topped black shoes, and a wide-brimmed hat, smiling mischievously.

Then came the portraits of Indji in her arrogant years. As she grew up, her dresses became longer and more opulent and even as a teenager, she still wore only white. In one photograph, she sat on a thronelike chair, her hair swept up in a pompadour. In 1905,

Indji Cattaui as a young girl in Cairo, dressed as always in white.

as she turned eighteen, a French artist was commissioned to paint her portrait. She posed standing against a ledge wearing a flowing gown made entirely of white lace and muslin, holding in her hand a single rose. She looked like the classic Edwardian beauty, delicate and dreamy.

Before the painting was finished, Indji Cattaui fell ill with typhoid fever. It was the curse of Egypt—*la maladie du pays*. An insidious disease that was rampant in the hot summers, it was as easy to catch as the flu. Because of the primitive public health measures, there were constant outbreaks of the dreaded *fievre Typhoïde;* and in those days before antibiotics it was almost impossible to treat—the fever had to take its relentless course and some survived but many did not. The pasha's daughter died before her nineteenth birthday and was buried in the family mausoleum.

Indji Cattaui was destined to remain as it were frozen in her coming-of-age portrait—forever eighteen. And that was the burden the pasha's wife carried, the pain that a thousand soirees and sets of monogrammed china couldn't lessen.

Madame Cattaui Pasha wasn't to be seen crying in public—not ever. Her mourning was silent and private and eternal. The painting remained on display in the library of Villa Cattaui. When her oldest son and his wife had a daughter, they named her Indji to honor the young woman but also to console the inconsolable pasha's wife.

My mother, Edith, who came to know her as a young teacher in Cairo when she worked at L'École Cattaui, and loved her with all her heart—loved her like a daughter—believed Alice Cattaui to be the most formidable woman in all of Egypt. Yet even my mom who became so close with her never had an inkling of the pain that lurked within.

Every few months, Madame Cattaui would get into her chauffeured limousine and quietly instruct the driver to take her to the small family cemetery where she would disappear to linger at her daughter's grave. She then would return to the car for the drive to Garden City or to work at the palace at Abdeen or the palace at Heliopolis or the palace at Koubeh.

*I*n 1936, King Fouad died suddenly, and the pasha's wife was left once again bereft—he had been her mentor and greatest patron. But with the same force of personality that had made her essential to Fouad, she became indispensable to his teenage son Farouk, who she had known since he was a baby. She was there when Farouk took power that spring, and she was a witness and intimate participant at his coronation the following year, when

King Farouk at his coronation, 1937.

more than two million Egyptians flocked to Cairo to watch the seventeen-year-old monarch ascend to the throne.

Unlike his late father, Farouk was fluent in Arabic and dazzled the crowd when he delivered his first speech in their native tongue.

When Farouk was married two years later, Madame Cattaui became lady-in-waiting to his new bride, Queen Farida. She even emerged as a matchmaker of sorts for Farouk's sister and the future shah of Iran. Back in 1931, at a party given by the Iranian delegation, the pasha's wife had paused in front of a painting of the young Reza Pahlavi and remarked: "Who is this handsome young man?"

That was enough to give an Iranian minister an idea: to bring together the two great families of Egypt and Persia. A few years later, the young Pahlavi became engaged to Princess Fawziah, the

The pasha and his wife at their fiftieth wedding anniversary, Cairo, 1930s. Portrait taken by the renowned photographer, Jean Weinberg.

most beautiful of Farouk's sisters. The minister credited Madame Cattaui for the union.

It was such a glittering period. One grand occasion followed another, none more elegant and festive perhaps than when the pasha and his wife marked their fiftieth wedding anniversary—*c'etait leurs noces d'or*—with a high tea thrown in their honor at Villa Cattaui. Jean Weinberg, the legendary photographer to the Royal Court, took pictures of the occasion. What he produced was vintage Weinberg—a portrait of the pasha and his wife that captured the soul of its subjects.

Yussef and Alice sit regally side by side on a velvet couch. Traces of their sumptuous lifestyle are evident in the carefully assembled frame—the rich embroidered pillows, the elaborate Oriental rugs, the satin and veil window curtains that part dramatically behind the couple.

The pasha looks magisterial in his *tarboush,* or fez. He holds a cane in front of him between his open legs, in the manner of a typical Middle Eastern gentleman. Alice is every bit as formidable in her flowing dress with an embroidered collar, her signature pearls, and the white gloves she clutches tightly in her hands. On her head is a jaunty cap with ostrich plumes. But while the pasha appears content, a hint of a smile on his face, a glint in his eyes, his wife, the arbiter of Egyptian society, stares straight ahead, serious and sullen and utterly mournful.

2

The Alley of
the Pretty One

*T*here was that blissful hour every afternoon when residents of the Alley of the Pretty One would venture out to their balconies, sit back, and savor their *café turc*, and the narrow little lane turned festive and joyous. Families put the difficulties of the day behind them. Friends and even strangers strolling in the area were entreated to stop by. Everywhere, there were cries of *Ahlan musahlan*, the effusive Arabic greeting that manages to say hello and welcome and please come join us in two words.

The ritual of the late afternoon Turkish coffee was sacred throughout Sakakini, a cozy neighborhood where many of Cairo's Jews lived in small buildings they shared peaceably with their Muslim and Coptic neighbors. Sakakini's streets were lined with four- or five-story walk-ups; the taller, more stately structures, a few with elevators, could only be found on the adjoining boulevards such as Malaka Nazli. Finally, there were the little serpentine alleyways, dozens and dozens of them, where the residences were ever so humble, made of ancient stone and typically only a couple of stories high.

There, the roads were dusty and unpaved, and even one car could barely get through.

Haret el-Helwa, the Alley of the Pretty One, where my mother, Edith, lived with my grandmother Alexandra, was longer than most alleyways and a bit more distinguished as it had its own mosque, smack in the middle. Five times a day residents could hear the imam's stirring call to prayer, "Allahu akbar."

Every apartment had a balcony, and in those days when even a ceiling fan was a luxury, it became an essential gathering place come dusk, when the intense heat of Cairo finally broke and a wonderfully refreshing breeze would drift in from the Sahara.

That is when the balconies would fill up with people—grandparents, rambunctious children, tired housewives—and they'd bring out serving tables crammed with snacks to nibble on, *à grignoter,* such as freshly peeled cucumbers, roasted pistachios, slices of mandarin oranges, or sticks of sugar cane that everyone young and old loved because they were so juicy and delicious. Occupying the place of honor was the *tanaka,* the little copper pot with the long handle filled with steaming Turkish coffee, sweetened for good luck.

It was a ritual enjoyed by all. The only difference was that the Jews tended to speak French among themselves, whereas Muslims conversed mostly in Arabic and the Copts a mixture of the two. But it was a sign of how well everyone got along that in addition to the mosque there were at least four major synagogues within a few blocks.

Whenever a cool breeze would blow in, you could hear the

murmurs of relief. "Enfin, de la tarawa"—finally, some fresh air—someone would cry in that mixture of French and Arabic so many favored in this culture that managed to be both European and Middle Eastern.

Most of the alley's residents were content simply to be outside, enjoying the street life. Even in the evening, vendors were still on the prowl, and you could count on seeing *le marchand de robabekiah*, the merchant of used wares, pushing his little cart filled with the oddest odds and ends—a broken, naked doll, a rusty cooking pot, a cardboard box. He'd come by chanting "Bekiah, bekiah" ("Old junk, old junk"), hopeful even at this late hour to make a deal of a piaster or two that would redeem his day and make his hard labors walking in the heat worthwhile.

My grandmother Alexandra, who was fluent in Italian, would shake her head at their ignorance, the fact that they didn't even realize they were citing an Italian phrase—*roba vecchia*, old clothes—they had appropriated and made their own and turned into a colloquial Arabic expression.

Alexandra and Edith were fixtures on their balcony, joined occasionally by Félix, Mom's younger brother. They lived on the ground floor—the least expensive dwelling in any building—and even paying for that was a struggle, and most of the neighbors knew it because everyone knew everyone else's business.

The apartment was small and dark—not much sun ever penetrated the narrow alleyway—and sparsely furnished. The living room in the front had only *le canapé*, a small couch that doubled as a bed for Félix. My mother and grandmother shared the bedroom in the back. And then there was Alexandra's piano, the most essential object in the house and the most incongruent, which she played less and less as she grew older.

Edith was the Belle of the Alleyway, and some liked to say she had given the alley its name. That was fanciful, of course: Haret el-Helwa had preceded the young girl by decades if not centuries. Still, like a lovely fable, a myth that springs out of thin air and takes hold, the story was repeated and spread.

Her beauty didn't give her too many advantages. Edith lacked both male and female companions, except for my grandmother who wouldn't let her out of her sight. Alexandra was prone to melancholy, and the only person who could reassure her and keep her calm was her daughter. The two were constantly together—to be separated even for a few hours of a day was unbearable, certainly to Alexandra.

They made a striking pair. Alexandra, small, bent over, pencil thin from years of barely eating, was always slightly unkempt, her gray hair awkwardly pushed back in a loose chignon, her clothes worn to the point of shabbiness, the buttons of her sweater missing or slightly askew, her stockings drooping or with holes in them, her shoes frayed and dusty, with a book or magazine tucked under her arm. Edith, on the other hand, dressed impeccably, in fitted skirts and delicate blouses and high heels. With her perfect posture, she actually appeared taller than her five-foot frame, and she had an innate sense of style that made her stand out among the women of the neighborhood.

Alexandra took immense pleasure in her daughter's beauty, and she loved to say that Edith couldn't walk down a street in Cairo, even in the fashionable quarters *en ville* (downtown), without drawing stares.

In 1937, shortly after her fifteenth birthday, my mother landed her first job—a position as a schoolteacher at Le Sebil, the communal Jewish school that was heavily supported by the Cattaui family. It was a remarkable achievement, not simply because of her youth. Teachers were a kind of aristocracy, and they were treated with absolute deference not only by schoolchildren but by other adults. Women who entered the profession were held in especially high regard in an era when girls were pushed to do little more than get married as young as possible.

And that is exactly what most of them did.

Despite the flowering feminist movement unleashed the prior decade by Hoda Shaarawi, Cairo in the 1930s didn't encourage a woman to get an education, and girls rarely remained in school beyond the age of twelve or thirteen. Many of the young Jewish girls were snapped up by the local *couturieres*, or seamstresses, to work as apprentices and learn the craft for a few years before settling down.

Those who were pretty and worldly found slightly better jobs at the big opulent department stores. Cicurel, the grandest of them all, placed a high premium on beauty and tended to hire strikingly attractive young Levantine girls barely in their teens for their sales force. They were paid both a salary and a commission, and the more aggressive among them did quite well.

Edith could never see herself as a shopgirl, nor would Alexandra have allowed it. Instead, she stayed in school an extra couple of years until she earned the coveted *brevet*, a kind of high school degree, and along with it, a license that allowed her to teach. Few women in the community ever went as far as obtaining the *brevet*, and Mom considered it her single most precious possession on earth. She didn't let it out of her sight and never would. Dated May 1937, it was issued by the Academie de Paris, the august body in Paris that set educational standards for schools all over France. The Academie was also in charge of overseeing French schools abroad that followed their curriculum. "Mademoiselle Matalon has been judged worthy of receiving a Brevet to teach elementary school," the large brown document stated.

There she was, working in the same school the pasha's wife loved to patronize. My mother had noticed her, of course, as she swept into the school to do her good works in the cafeteria or with the needy students. But she was too shy to approach her and could only marvel at her from afar, admiring both her elegance and the tenderness she lavished on the children.

Situated in a massive brick building, Le Sebil had classes of forty students and more. In addition to the Cattauis, other moneyed families also helped—the Jews who lived on the other side of Cairo—in the gated villas of Heliopolis and Zamalek and Garden City and Maadi. While these didn't necessarily mix with the Jews of Sakakini or the Old Ghetto, they were committed to making sure the less fortunate in the community had enough food to eat while their children received a solid education.

Academic standards were fairly rigorous, but life at the Sebil was grim. My mother was shaken by the extent of the poverty among the children. Many came to class in the morning hungry, and she was painfully reminded of herself as a little girl, foraging for food with Alexandra. "Ce sont des pauvres hères," she'd tell her mother sorrowfully; They are all these poor ragamuffins.

In the absence of a government welfare system, the Jews of Egypt tried to take care of their own with a far-flung network of charitable funds set up to help every needy group imaginable— orphans, widows, the sick, the aged, the destitute, the insane. There was even a pot of money for single women in danger of never marrying because their families couldn't afford a dowry. It was a favorite cause of the pasha's wife, for whom every bride was Indji, every dowry she subsidized was one she would have arranged for her dead daughter.

In Edith's case, her slender means didn't seem to matter. Matchmakers were already circling; they had approached my grandmother about potential suitors anxious for her daughter's hand. These were said to be so smitten that one or two were prepared to forgo *la dotte*—the all-important dowry.

Alexandra wouldn't hear of it. She made it clear she wasn't

going to turn her daughter over to *le premier chien coiffé*, the first well-groomed dog, as the French liked to say, the first rake who came along. She was sure that Edith could make a dazzling match someday, and she was still so young they could afford to wait. Besides, it would be so hard living apart—now that her daughter went to work every morning, the day somehow felt longer, more difficult to navigate.

Edith left early for her teaching duties at the Sebil, and Alexandra was left to fend for herself.

My grandmother was so terrified of being alone she'd simply remain on the balcony and wait for Edith to come home. She'd be fixing herself coffee, boiling the water with several spoons of the thick Turkish blend; and for nourishment she nibbled on some *khak*, the salty, ring-shaped biscuits sprinkled with sesame. She was also constantly lighting up—one cigarette after another she bought from urchins on the street. These were young boys who made a living gathering discarded cigarette butts that they turned over to slightly older youths who specialized in removing any precious leftover tobacco from the stubs, which they'd then recycle. There was a whole industry in Egypt rolling these primitive, cheap cigarettes that were sold loose on the streets, and Alexandra was a devoted customer.

She'd plunge into some novel her daughter had managed to filch for her because in the same way that my grandmother was always smoking one cigarette after another, and drinking one cup of coffee after another, she couldn't survive without reading one book after the other.

When Edith finally came home, the two women would start the evening side by side on the balcony, sipping coffee and chatting. Later, they'd take a stroll over to Sakakini Palace around the corner. Years after it was built, it was still an object of wonder, this fantastical mansion with its domes and golden statuettes and turrets. There it stood in the middle of a vast traffic circle from which eight different streets fanned out, like spokes of a wheel. The palace not only anchored the neighborhood, but gave it a certain cachet and even its name.

In their evening walks, Alexandra, arm in arm with Edith, would cross over to the palace. Often, they continued their stroll and went to pay a call on Alexandra's stepdaughter, Rosée, who lived nearby. They could sit with her and talk for hours even as she plied them with coffee and food. On a hopeless quest to persuade both women to eat, she'd serve one dish after another. Sometimes, Edouard—Edith's favorite uncle—would be there. Edouard was a dashing fellow who had clawed his way out of poverty and now lived downtown, where he thrived as a pharmaceutical salesman. He had a special fondness for Edith, and so when he was there, the atmosphere was especially joyful.

Later, if they had a bit of change, my grandmother and my mom would catch a double feature at the *cinéma en plein air,* the small outdoor movie theater near Sakakini Palace whose nightly show started promptly at 9:00 P.M. The cinema offered comfortable bamboo chairs where every evening, spectators could lean back and enjoy the latest Hollywood movie while eating a sesame roll with cheese from the vendors who walked up and down the aisles; and my grandmother, who loved the movies almost as much as books, was finally at peace.

Typically, the shows lasted till midnight, and when they came out, the neighborhood was still awake and alive, and they could buy a bag of roasted chestnuts from one of the vendors stationed outside the cinema and nibble on them as they strolled home.

Those evening hours—the *café turc* with Edith on the balcony, the pleasant exchanges with neighbors, the little constitutional walks, the occasional double feature—were so precious to Alexandra. It was hard to be a woman on your own in Egypt even if you had relatives or friends. People always looked down on you, felt sorry for you, avoided you.

Edith was her life. Alexandra knew that she could rely only on her daughter who was so much older than her years. Edith the diligent one, Edith the studious one, Edith the pretty one, Edith the dark-haired, doe-eyed hope of the family, the one who would rebuild their lost hearth.

How they had survived after her husband abandoned them, Alexandra was never sure. The recent years in the little ground-floor apartment were the good times. In the late 1920s, when Edith and Félix were children, Alexandra's husband, Isaac, had squandered what money they had, then left the family to fend for themselves. My aristocratic grandmother—cosseted and spoiled as the child of wealthy parents—hadn't known how to cope.

Even as she watched Alexandra unravel, young Edith took charge and looked after the three of them, preparing small elemental meals with the bits of food she could gather, keeping the house clean, caring for Félix. The passionate attachment between mother and daughter dated back to that period, when Alexandra had only the most tenuous grasp on reality and Edith was trying to save her with all the might and determination and ferocity that a strong-willed little girl can summon. She was forced to grow up fast, to forgo any childhood pleasures.

To forgo any childhood, period.

They had practically starved, and they couldn't afford any rents at all, even in the dustiest alleyway, even in the Haret-el-Yahood, the ancient Jewish ghetto where the desperately poor lived in the equivalent of rabbit warrens. So off they went to Alexandria, where my grandmother still had relatives who remembered her, who loved her, who could be counted on to give them a place to stay.

For a time the three of them—Alexandra, Edith, and young Félix—were nomads seeking refuge with various aunts or cousins. Their favorite, Tante Farida, Alexandra's half sister, took them in for weeks at a stretch. She treated Edith like a daughter and gave the family their own room. There were other stray cousins with whom they could dine so they were assured at least a square meal or two each day. But there were no offers of permanent shelter, and besides, my grandmother would never have agreed. She had trained Edith even as a child not to convey how desperate they were, and above all not to reveal that they were hungry.

Once a month they'd go off together on the most humiliating mission of all: to collect money from Alexandra's hopelessly cold relatives, those members of the famed Dana family who were almost obscenely prosperous and lived in villas and stylish apartments between Alexandria and Cairo.

They took turns giving my grandmother one Egyptian pound for her troubles—*une livre par mois*—that was her allowance. One pound. It was supposed to cover her expenses raising the children, feeding them, and keeping a roof over their heads.

Alexandra would pocket the one-pound note, take Edith by the hand, and return to wherever they were staying or travel back to Cairo. There, she would descend on other relatives, like her brother Edgar who also lived in Sakakini, and make an appeal. My grandmother was always so shy and apprehensive when she went to his house. She would practically tiptoe inside, sit quietly in an armchair, and not budge, hoping Edgar would give her some alms.

That was how they had lived for years and years—at the mercy of relatives who weren't especially merciful.

Despite all the drama and uncertainty at home, Edith threw

herself into her schoolwork. She was a star pupil of Marie Suarez, the girl's division of L'École Cattaui. The newly built private school had opened shortly after she was born and quickly acquired a reputation as the finest Jewish private school. Alexandra had refused to send her to the Sebil—had simply put her foot down. The notion of Edith attending a communal school for the poor was unthinkable to her despite her own penury, and she appealed to her relatives to pay the school's tuition so that Edith would receive a proper education. My mother was constantly reading and studying and earning accolades. She skipped several grades, received *tous les premiers prix*—all the top prizes—and was ahead of all her classmates, including those who were considerably older.

From its founding in the 1920s by Moise Cattaui Pasha, then the head of the Jewish community, the school had tried to recruit the finest teachers; and Edith reveled in their love and the attention they lavished on her. The teachers in turn raved about her to Alexandra. What a beautiful penmanship her daughter had—each letter, every word, so perfectly formed. And her compositions were so thoughtful and lyrical.

This was at least in part my grandmother's doing. Unable to offer Edith any material possessions, too distraught to give her the modicum amount of stability she needed growing up, Alexandra could only infuse her daughter with her own literary sensibility—her passion for books—along with a boundless, all-consuming, near-hysterical love.

There were times Edith chafed at that love and tried to keep my grandmother at a distance. One year that my mother was set to collect all the awards, Alexandra was invited to attend the ceremony. My grandmother decided to wear the lone "dressy dress" she owned, a deep purple velvet outfit she called in her typically fanciful manner, "ma robe de velours héliotrope"—my heliotrope-tinted velvet dress. It was from another era and hopelessly inappropriate for this hot June day. As they started walking toward school, my mom realized that they made a ridiculous pair and told Alexandra to go home—that she was embarrassed to be seen with her. My grandmother, looking like a wounded bird, walked back to the alleyway alone in her heliotrope dress while Edith went on to collect her prizes.

Usually, they got along and were devoted to each other and exceptionally close. From a young age, Edith was reading the same novels as her bibliophile mother—adult novels, novels intended for someone older and more mature. My mother had the most sheltered upbringing imaginable, yet in some ways it was also the most progressive because of the books she read. She learned about love affairs while reading Flaubert's *Sentimental Education*. Her exposure to marriage and its disappointments and betrayals came from *Madame Bovary*.

When she turned fifteen, Alexandra handed her perhaps the greatest gift of all—Proust's *A la Recherche du Temps Perdu*. She had managed to obtain a complete set from an obliging cousin. Within a few months, my mother had read every single volume.

This became Edith's claim to fame—that she had devoured all of Proust by the age of fifteen. She would say this with a touch of arrogance, because, truth be told, she looked down on other girls her age who didn't read Proust, who had no interest in him.

There was no such superior streak in Alexandra, who'd been pummeled far too much by life. She was simply thrilled after the years of humiliations to be able to hold her head high again. These days, when she went to see her relatives, it was to brag a little bit, to give them all the wonderful news. She was overjoyed when her daughter received her teaching credentials. "Edith a reçu son brevet" (Edith got her diploma) she told anybody who'd listen. A bit later, she announced, "Edith enseigne" (Edith landed a teaching job). Yet even now that my mother brought home a steady paycheck, my grandmother still went quietly alone every month to collect her one Egyptian pound from balky family members.

As she grew older, Alexandra became more and more fearful. Along with her angst, or perhaps because of it, she was increasingly superstitious. Life was all about good luck and bad luck, how to encourage one and ward off the other, and above all how to avoid what was known in Arabic and Hebrew as *ein arah*—the evil eye. These beliefs were widespread in old Cairo, and Jews who were educated and cultured were every bit as terrified of the evil eye as their impoverished Arab neighbors. But the fear was carried to an extreme in the ground-floor apartment of the Alley of the Pretty One.

My grandmother was said to be ill fated from the day she was born. From the start, family members spoke of a Curse of Alexandra, which they traced back to her parents, specifically to the day her father, Selim Dana, had married Rachel Dana, a much younger woman who happened to also be his niece. Such near-incestuous unions were tolerated in Egypt—barely—but there were many who were shocked, who shook their heads and said the match was sinister and unlawful in the eyes of God.

"It will bring them bad luck," people whispered. *Ca va porter malheur.*

It did. How else to account for the death of Rachel at the age of thirty, leaving behind not only Alexandra but three young sons?

The offspring would be the ones to suffer, those who opposed the union predicted, recalling the biblical injunction about the sins of the father: "For I am the Lord your God, a jealous God, Who visits the sins of the fathers upon the children to the third and fourth generations."

Whenever Alexandra experienced another tragedy, another

setback, people merely shrugged as if to say, "I told you so." My well-read, intellectual grandmother also believed in the curse; she was convinced that she was star crossed.

Alexandra was certain her daughter was especially vulnerable to the evil eye because of her beauty and now her teaching job. My grandmother had a thousand injunctions for Edith to chase away the demons and keep those who wished her ill at bay. She believed that running into a priest was unlucky and if my mother ever saw a clergyman coming, she was instructed to cross the street immediately. Now, Sakakini had a Christian girls' school operated by the Sisters of Sion, so there were many nuns in the neighborhood. The injunction did not pertain to them: Alexandra, educated at a convent school, had a soft spot for nuns and referred to them lovingly as *les* sisters.

Every week, my grandmother persuaded Edith to join her in a ritual designed to combat those who might wish them ill. Together, they would burn incense purchased in Old Cairo at the *souk el attarine*—the souk of the spice merchants. In stalls fragrant with the scent of exotic herbs and spices, you could find the men who specialized in selling *bakhour*—the strange and powerful incense they prepared themselves by mixing aloeswood and musk and sandalwood and clove and cinnamon and lavender and myrrh and mastic and a thousand other secret ingredients.

Alexandra would place the incense in a metal container on the floor and throw a match over it. Almost immediately, the small apartment would be heady with the sweet-smelling fumes. She and Edith would walk around the burning incense seven times— once for every day of the week—to chase away any bad luck and to help usher in good luck.

This was a popular practice throughout old Cairo. There were even "professional" women who came to your home and burned incense and drove away the evil eye for a living. Mostly elderly, they'd arrive bearing their small brass incensory and walk from room to room, shaking it this way and that. The scent of the *bakhour* lingered for days.

But what was different about the house in the alleyway was my grandmother's insistence on burning the incense herself, and the degree of passion she brought to the task, and the fact that she instilled in my mother at a young age such a terror of the evil eye, along with the absolute belief in the power of those fragrant, smoldering bits of powder to shield her from harm.

My mother blossomed as a schoolteacher; at last, she was enjoying the status and reputation she and Alexandra had craved. When she walked into class now, all the little children immediately rose to greet her.

Though shy by nature, she emerged as a formidable figure in the classroom. She had a reputation for enforcing strict standards even among her youngest charges. But she also gave of herself more than the typical teacher: If a pupil didn't grasp a subject, she worked with them in school and after school.

She even made "house calls"—visiting them at their home to tutor them.

But she also brought an abundance of charm to her lessons. She could keep her young male charges spellbound with her fantastical stories, and, of course, she was so pretty that even the littlest of little boys were in love with Mademoiselle Matalon.

And so were some girls. Young Sarah Naggar, who lived on Ibn Khaldoum Street, around the corner from the Alley of the Pretty One, loved to stand on her balcony every morning watching the flow of traffic. She was always monitoring the comings and goings, but her favorite treat was that moment when the lovely Mademoiselle Matalon would pass. She was fascinated by how the teacher carried herself, how her outfits were so dainty and re-fined—not at all the way other women in Sakakini dressed—and the child could only dream of the day she, too, would put on heels along with frilly white blouses and tailored skirts.

After a couple of years, Edith was handed a plum—the chance to teach at L'École Cattaui.

She was back on familiar territory. The tony private school attracted a different class of students—not rich exactly but fairly well-to-do. Their parents had to pay a monthly tuition. The little boys who attended Cattaui arrived every morning looking spar-kling and polished in their satiny black uniforms with the large white Peter Pan collar.

Mom always had a weakness for a nice uniform.

There was an intensity to the education at Cattaui that ap-pealed to her. Young children were required to study four or five languages—French, English, Hebrew, Arabic, and even Italian. They were constantly being tested and given impromptu pop quiz-zes and oral exams and *dictees*, the French equivalent of spelling tests. Forced to think on their feet, children learned how to tackle math problems without relying on pen and paper but by making the calculations in their heads.

In an effort to emulate the finest Parisian educations—Paris was always the ideal—a battery of teachers were hired who spe-cialized in every subject under the sun: the basics, of course, such as French grammar and language, which my mother taught, as well as English, mathematics, history, biology and the Bible. There were even courses in calligraphy.

Rome's cultural attaché in Cairo, a devotee of Mussolini, sent over an Italian teacher, Signorina Messa-Daglia, to help out. She was a blond bombshell, tall, beautiful, and very Aryan looking, who promptly taught her classes the Fascist salute. A Fascist teacher in a Jewish school? Yes: After reciting the Jewish morning prayer in a communal hall, the Signorina's students greeted her by shouting "Viva Il Duce"—Long live Il Duce—with their arms outstretched.

But that was the only element of fascism that touched Cairo's Jewish children. Unlike their peers in so much of Europe in the late 1930s, they weren't subjected to horrific bouts of anti-Semitism or forced to obey demeaning racial laws.

On the contrary, from the royal palace where Madame Cattaui was establishing herself in the court of King Farouk, to the humblest alley in the Old Ghetto, Jews at every level of Egyptian society were enjoying unprecedented freedoms and opportunities. Their lives in this Muslim city were filled with possibilities in ways that had ceased to be the case in most of Europe.

Thrilled with her new position, Edith would get up early each morning, dress carefully with only a hint of makeup to make herself appear older than her seventeen years, and make her way over to Cattaui, a short walk from the alleyway.

The headmaster, a formidable figure named Monsieur Moline who ran both Cattaui and Le Sebil, took her under his wing. As part of his grand plans to raise academic standards and make Cattaui competitive with the Catholic schools of Cairo, which were said to be more rigorous, he was recruiting the best teachers he could find and was said to have a knack at spotting talent. He was a fan of the lovely and genteel Mademoiselle Matalon, gave her choice teaching assignments, and treated her with extraordinary deference.

He also introduced her to the school's most important benefactress, Madame Alice Cattaui Pasha. Once she became known to the pasha's wife, my mother—an outsider for so long— was finally on the inside; she had found a second home at 17 Rue Sakakini.

adame Cattaui Pasha was nearly seventy when my mother came to know her—graceful and proper and dignified yet with a touch of vanity, a hint of *la coquette* that prompted her to always wear a large choker or kerchief around her neck to hide any telltale signs of age.

She was very much the *grande dame* in her couturier clothes and white gloves and Paris hats. Yet she was neither cold nor supercilious. Indeed, her greatness lay in the fact that she could consort as easily with a king as with a child in need of a pair of sandals.

She also had a well of goodness within her.

A keen judge of character, she embraced Edith, who found herself the unlikely protégé of this powerful and stately woman. The pasha's wife became her patron saint and guardian angel and surrogate mother all at once. As a frequent visitor to the school, she could follow my mom's teaching career and recommend her for new responsibilities. And because this noblewoman had taken such a strong liking to her, the young teacher found herself elevated in everyone's eyes.

Edith's newfound stature was almost enough to make up for all the hardships and deprivations of living with my fragile grandmother. Her relationship with Alice Cattaui Pasha made Edith intensely proud, and that pride lingered through the years and decades, sustaining her even when all semblance of pride was gone.

Whenever Madame Cattaui came to the school, the two would huddle, conferring intently on the questions of the day. Edith clung to her as if she were her mother—a mother with the strength and stability that poor Alexandra could never muster. And Madame Cattaui returned the young woman's affection, treating her like a surrogate daughter—like the daughter she had lost. Because surely that is what happened: My mother was Indji, returned to life again in the form of a delicate and exquisite schoolteacher.

The friendship came with privileges. My mother learned all about the pasha, Yussef Cattaui, and his extraordinary and boundless love of literature, the thousands upon thousands of volumes he kept in the private library at Villa Cattaui—every single possible work of note.

Would she like to borrow some?

It was a startling offer—deeply affecting to my mother who'd had to scrimp and save to afford a steady supply of books for herself and Alexandra over the years; books were so expensive in Cairo. They'd always relied on friends or cousins to give them access to the works they craved or bought them secondhand.

One day, after school, Edith made the journey by tramway from Sakakini to Garden City. It was like traveling to a foreign country—instead of bustling alleyways, there were quiet, landscaped streets, with homes larger than any she had ever seen, with the exception of Sakakini Palace. Finally, she reached 8 Ibrahim Pasha and was ushered into the wing that housed the sumptuous Bibliotheque Cattaui.

With its vaulted ceilings and stained-glass windows, the pasha's library looked like a church—a cathedral of books. On shelves that lined the walls, many shielded by glass, there were all the authors she knew and loved as well as some she had only heard about and longed to read. She glimpsed a set of Flaubert's *Oeuvres Complètes*, from *Madame Bovary* to the *Temptation of St. Anthony*. Every novel and poetry collection by Victor Hugo was there, including *Notre Dame de Paris* (*The Hunchback of Notre Dame*) and *Les Miserables*, both childhood favorites. There was Emily Brontë's *Les Hauts de Hurle-Vent* (*Wuthering Heights*) and Tolstoy's *La Guerre et La Paix* (*War and Peace*) and assorted works of Pascal and Emil Zola and Guy de Maupassant and the Goncourt Brothers, all in first editions. Each had an elegant label pasted inside, with the Cattaui family monogram. And, of course, there was Proust—multiple editions of his works containing every volume he had ever published—in jackets of burnished leather.

The pasha's collection was maintained by a librarian who lived in the Cattaui residence. He was courtly and polite and eager to give Edith access to any works she requested. She settled on *Les Thibaud*, a French *War and Peace* that was all the rage. The sprawling family epic by Roger Martin du Gare had recently snared the

The pasha's library.

Nobel Prize for Literature. She lugged all eight volumes of *Les Thibaud*, from Garden City to the alleyway.

Once Alice realized my mother's passion for books, the bond between the two women only deepened. Madame Cattaui became sensitive to a widespread problem within the Jewish community—the difficulty for impoverished students to purchase the books required for their studies. Although she and her husband gave to countless philanthropic funds and communal institutions, she had never focused on the struggle families faced to get hold of needed books until Edith told her about her own travails.

L'École Cattaui would change all that. Madame Cattaui decreed the school would have a state-of-the-art library and *la chère Mademoiselle Matalon* would be the one to organize it. That was the extraordinary project entrusted to Edith—to set up a library that would allow even students of modest means, or simply those with great intellectual curiosity, to read and study and take home any books they fancied.

The two women worked closely together, and my mother was given a budget to order any and all works she felt were important—novels, history and geography texts, biographies, whatever could be of interest to young students. She could purchase them from booksellers as far away as Europe and America; they all knew the Cattauis—the pasha had done business with each and every one of them.

It was a thrilling assignment, and Edith, still in her teens, rose to the challenge. Driven, committed, and thoroughly impassioned by her undertaking, she had never felt so empowered as when she ordered more books, and money was no object, and she could indulge in all her tastes.

She performed her duties with such flair that her reward for creating the library came in the form of a new position. Madame Cattaui installed her as the school's librarian. She now had two titles and two jobs, since giving up teaching her beloved little boys was out of the question.

Whether she was in the classroom or ensconced in her fledgling *bibliothèque* or basking in the companionship of the pasha's wife, Edith had never looked as radiant or appeared so self-confident. She had reached her arrogant years, that period in a young woman's life when she feels—and is—on top of the world. Although it was a small world, bounded by Sakakini Palace and the careworn Alley of the Pretty One and the Cattaui school, to Mom, it suddenly seemed infinitely grand.

And perhaps only Alexandra, only my poor superstitious grandmother, trembled at her daughter's sudden surfeit of joy and good fortune and wondered what evil wind might sweep in to take it all away. She begged her daughter—as Mom would one day entreat me—to be mindful of the *mauvais oeil*, those malevolent forces that lurked everywhere around a person who seemed to have too much.

But the rewards kept coming, and my mother's luck defied Alexandra's dark misgivings. Mom continued teaching, ministering to the mischievous boys of *la maternelle*, who were in kindergarten and first grade, and then, her duties in the classroom over, she would turn her attention to the library.

One day out of the blue, Madame Cattaui arrived with a gift. She handed Mom a key—the key to the pasha's library.

My mother would always evoke that moment when Alice Cattaui Pasha placed the key in her hands. As a little girl, I would watch mesmerized as she acted out this scene as if from a long-ago play, recalling what happened in a dreamlike voice and obsessive precision:

"La clef, Madame Cattaui Pasha m'a donné la clef, elle l'a mis

Edith on her wedding day, Cairo, 1943.

dans ma main"—the key, Madame Cattaui Pasha gave me the key, she put it in my hand, she would tell me over and over again.

For Mom, the key was a precious and ultimately transformative gift that offered a different way of looking at herself forevermore. She was no longer simply a girl from an alleyway, but the fine and talented Mademoiselle Matalon, beloved by a pasha's wife, embraced by one of the most powerful women of Cairo.

3

The Bride Who
Set Herself on Fire

Alexandra had never looked so radiant as that afternoon when she rushed into her brother's house to announce her beloved Edith had encountered "quelqu'un de très grand," someone of great stature, high up in the world. He seemed thoroughly smitten and there was suddenly no time to waste: He wanted to meet the family immediately and announce the engagement.

The words came tumbling out of my grandmother. Edith had captured the *gros lot*—grand prize in the lottery—as this gentleman seemed to have every quality one could possibly want in a husband: wealth, social pedigree, impeccable manners, and, of course, he dressed so wonderfully, with such style.

"C'est un vrai aristocrate," she exclaimed, unable to contain herself—He is a true aristocrat.

Her relatives, who had long feared that my grandmother would let her daughter become an old maid rather than part with her, were mystified, at a loss as to how to explain this sudden change. Alexandra had always been such a fierce guardian of Edith—for years, no man could ever hope to approach the young woman who was watched more closely than the crown jewels. How many matchmakers had been spurned for daring to suggest a prospect she didn't think was worthy enough. From the time Edith was fourteen or fifteen, my grandmother had been fending off potential suitors as unimpressive, not educated enough, not wealthy enough, or simply lacking personality. She had even stopped Edith from becoming involved with the one young man who came closest to capturing my mother's fancy—an *étalagiste*—a charming young window dresser who worked at the great department stores downtown. He had told her she belonged in one of his windows.

Anyone who knew Alexandra—who saw how she clung to Edith—suspected she wanted to keep her daughter always by her side on that little balcony, sipping cup after cup of *café turc* through eternity.

Now here she was with Edgar and his wife muttering that an engagement—*des fiançailles*—was close at hand. My grandmother kept talking about Shepheard's, the hotel favored by British officers and anyone else from the swell set. It was the spring of 1943—the middle of the war—and the Brits were ensconced in every inch of the hotel's famous bar and veranda. The plan was to meet the prospective groom for high tea.

"But why don't you come here to our house," Marie, Edgar's wife, told her. She was perfectly prepared to host a luncheon and have everyone over. "Qu'il vient chez nous," she said—Let him come to us.

Alexandra, usually so weak and timid, was unusually firm. Absolutely not, she replied—the gentleman, his name was Leon, was insisting on Shepheard's, and my grandmother was clearly impressed that her future son-in-law wanted to host a get-together in the swankiest corner of Cairo.

Shepheard's on Ibrahim Pasha Street was an old dowager of a hotel, fusty and on the careworn side. Yet among the Brits and foreigners who descended on Cairo, anyone who was anyone favored it above all others, including Winston Churchill who would only eat meals prepared by the hotel chef when he met secretly with President Roosevelt. Another fan was the Nazi commander Field Marshal Erwin Rommel. The previous year, when Hitler's troops were poised a heartbeat away from Alexandria and the British seemed on the verge of losing the battle at El Alamein and with it, the entire Middle East, rumor was that the German "Desert Fox" had gone ahead and reserved a suite of rooms at Shepheard's. That way, the world would know that Germany had defeated Britain not simply on its battlegrounds but in its barrooms and its tea salons.

In her walks around the city, Alexandra had never dared enter the hotel; she could only stare in her usual distracted way, taking in the British officers sitting on the comfortable bamboo chairs on the terrace and looking so imperious as they were served by one of the silent, intensely elegant Nubian waiters in their flowing white robes with a red sash and matching red *tarboush*. It was a man's world, that terrace. The women tended to congregate more discreetly indoors, in the parlors and dining rooms. Watching all the Rolls-Royces pulling up one after the other—including on occasion King Farouk's red cars—was a full-fledged spectacle, like going to the cinema.

As the war raged, it was one of the few places in Cairo that still served first-class champagne.

And that was where Alexandra and Edith and Félix were headed that afternoon when they boarded the tramway on Sakakini Street, their relatives having gleaned a bit more information over the intervening days from my grandmother. She had filled in the gaps in her usual anxious, telegraphic style.

A chance encounter at La Parisiana, the popular café downtown. A handsome man in white—white sharkskin—had made their acquaintance. *Très distingué* (terribly distinguished). He had sent a note over to their table. *"I find you very beautiful—Would it be possible for us to meet?"* His name was Leon. He was much older than Edith. A businessman.

"What kind of business?" Alexandra hadn't the foggiest. How old was he exactly? She didn't know that, either. Those were all petty details, and my grandmother was never ever able to size up real-life men and women the way she expertly judged the characters in the books that she devoured.

"Ça sera un marriage de rêves," she told her brother and his wife as she left. It will be a dream wedding.

When Edgar and Marie, dressed in their afternoon finest, arrived at Shepheard's, they were swept up in a world of tall granite pillars, Persian rugs, stained glass, and the waiters, always the waiters, scurrying past them so silently in their slippers they seemed to float through the rarefied air of the hotel instead of walking like ordinary humans.

My father usually favored the American Bar or else the Long Bar. Its legendary waiter Joe knew so many people in Egypt everyone assumed he was a spy. But for this occasion only a cozy parlor would do. About a dozen relatives gathered at the appointed time for tea, a perfectly civilized way for the two families to get to know one another. My father was there with several of his siblings and their spouses. Edith arrived with Rosée, her half sister, who was almost like a mother to her, and Edouard, her half brother, who was also very protective. Edouard was the most worldly member of the family, the only one who felt at ease in Shepheard's. Edouard boosted my mom's confidence as no one else could, giving her the sense that she was equal to anyone and everyone in this opulent Orientalist drawing room. He was there to remind her that she needn't be intimidated.

When Marie, my grandmother's sister-in-law, saw Leon, she did a double take. Though she hadn't remembered his name, she recognized him at once. Marie was a veteran poker player, a habitué of card games all over Cairo. A passionate gambler, she was one of those women who felt perfectly at ease playing with—and beating—a table full of men, whether at poker, baccarat, or rami, Egypt's version of gin rummy. Card games were one of the few intellectual outlets for women in the Levant, and Marie approached every round with a zeal and ferocity that surpassed that of the men.

The two had a startling reunion: "Mais qu'est-ce-que tu fais la?" she asked Leon. What are you doing here?

"Je suis le fiancé," he replied—I am the fiancé.

Marie was genuinely stunned. In her years playing cards, she knew Leon to be *un coureur,* someone fond of the ladies. Cairo was both an international city and a small town—everyone knew everyone, everyone was constantly running into everyone. It was also a culture that thrived on gossip. If you happened to be on the "night circuit"—going to the same favored set of nightclubs, casinos, and cabarets, playing at the same popular poker tables—it didn't matter if you were a gambler or a king, people watched you, observed you, noticed who was with you and who wasn't.

As members of both families sipped tea and traded niceties and wished each other well, Marie found herself that afternoon trying to solve two mysteries at once: why her old poker buddy had decided to marry, and why her sister-in-law was allowing Edith to be his bride.

After the Shepheard's tea, Marie decided to confront Alexandra with what she knew. Leon wasn't a suitable prospect for her daughter, she told her—he was known to run around and he was notorious for dating *des actrices*—actresses. That was a short-hand way to convey the kind of women my father was famous for squiring, women who weren't remotely like Edith—experienced, which my mother was not, and free, which my mother was not, and worldly, which my mother most certainly was not.

"Non—ce n'est pas vrai," my grandmother insisted when Marie had finished—It is not true.

Alexandra, who had always lived in her own world, poised somewhere between her favorite novels and her most cherished fantasies, had constructed a new reality for herself, one featuring a noble and wealthy son-in-law.

Nothing her blunt-spoken relative could say would dissuade her. She and Edith had waited too long, had prayed too intently, and the answer to their prayers had come in the form of this tall dashing man in white. In her excitement, Alexandra had forgotten her own disastrous past.

And what did Edith think? She was in a state of confusion—swept away by my father and his grand courtship, her first, since Alexandra hadn't permitted any others. She was also heavily under my grandmother's influence—and Alexandra seemed determined, obsessed even, that this match take place.

Edith was twenty; Leon was forty-two. In that sense Alexandra was turning out to be very similar to other parents of the era, intent on marrying off their daughters as fast as possible. Rich or poor, girls were considered a burden to their families and an aging, unmarried daughter could only bring dishonor. It was critical to find husbands who would support them.

It was the mind-set known as *Yalla.*

"Yalla," a typical parent would tell their daughter when a prospective suitor showed up at the door. *Yalla* was the Arabic expression that suggested impatience, the single word that meant come on, get going, make up your mind, hurry up already. Like a thousand other mothers around her, Alexandra was ready to give her daughter away, beguiled by the promise of money and stature from men who were far too old.

When it came to marriage, a young woman in Cairo was in the same boat whether she came from an alleyway in Sakakini or a grand villa in Heliopolis or Garden City—none had much to say in the matter. The decisions, down to what type of *bonbonnières*— silver or porcelain?—were going to be given to guests as keep-sakes, were made without her input, between her parents and the groom to be and his family. The young women were kept docile by the promise of a lovely wedding, an elegant trousseau, new clothes. Off they went, often in a state of abject terror, to their new lives.

Though marriages to older men were common, some young brides still stood out. Sixteen-year-old Camille Wahba greeted her new husband holding her dolls in her arms. There was also the thirteen-year-old girl who was married off to a Syrian Jew in his thirties, a pastry salesman from Aleppo who sold sweets in the street. Whenever they boarded the tramway together at Sakakini, she rode for free: The conductor assumed she was his child.

Many of these young Levantine brides found themselves trapped, unable to escape. They learned to make their peace or, in a few cases, they got out but only after paying an excruciating price for a divorce. The majority stayed put. They grew up fast, were mothers within the first year, and in some cases became fond of their older spouses.

Some unions had nightmarish denouements—none more so than that of Ninette Toussoun, a blond, blue-eyed beauty and per-fume salesgirl at Orozdi-Back, one of Cairo's great department stores. She married a coworker when she was only a teenager, but the marriage soured. Her husband liked to go out gambling and would leave his young wife night after night. Ninette was said to wander from one Cairo building to another, possibly having dalliances. When her husband threatened to leave her, she said she would kill herself, and then she did.

In an ultimate gesture of despair, Ninette threw oil over her body and set herself on fire in front of him, leaving her Jewish neighbors stunned and anguished and determined to make sense of what on earth had taken place. Did he have affairs? Did she? Had he been unkind? Had she gone mad? No one knew for sure. For years to come, the community couldn't stop talking about the burning of the Toussoun bride and tried to grasp what had led a young woman—one of their own, not from the Muslim community—to feel so unhappy that she preferred to burn to death.

None of these morbid prospects were on my mother's mind as she prepared for her wedding. On the contrary, she was as excited as Alexandra, swept up in the same grand illusion as her mother. Leon had wanted a fancy affair, black tie. He was willing to pay for the ceremony to be held at the Gates of Heaven and had arranged for the chief rabbi of Egypt, Rabbi Haim Nahum himself, to preside—which meant that he was going to pay a great deal.

There was only one sad note to the affair: Edith had to abandon her cherished position at L'École Cattaui. She would no longer be working as a teacher or as a librarian: Leon had made it clear she would have to stop working.

It was a condition not simply of this marriage but of nearly any marriage in Egypt. *C'était la loi du pays*—it was the law of the land—as people liked to say, the Oriental way, probably more reflective of an Arab than a European mentality. Once she had a husband to support her, a woman simply wasn't supposed to work.

And so while my mother had profound qualms about leaving a position she loved with all her heart and that had been not only her salvation but that of her family—in her five years as a teacher she had ably supported Alexandra as well as her little brother— it didn't even occur to her to challenge the need to abandon her profession. For all of her talents, my mother was a creature of her time. To resist my father's wish that she leave her job would mean going up against an entire culture.

Madame Cattaui was shocked by the news. As long as she had known her, Edith had been completely devoted to her work at the Cattaui school. Her young protégé had not expressed any interest in settling down, had never even had a romantic involvement of any kind. Yet there she was suddenly leaving to be married. Alice had lost a daughter once and now it was as if she were losing her all over again. But the pasha's wife was her usual composed, stoic self as she wished her well in her new life.

For Mom, the loss was far greater. Leaving Cattaui was akin to shedding her identity. She had been among colleagues who respected her despite her youth and inexperience, was idolized by any number of young children, and basked in the admiration and support of the magnificent pasha's wife. Yet she was leaving all that behind and more, for the promise of wealth and domestic stability that her new fiancé emblemized.

She wasn't asked to give back the key—that would surely have been unbearable. But with her departure from Cattaui she had effectively lost entry to the pasha's library and the entire magical world it conjured, and the promise it signified, and that fact alone cast a shadow over a period that should only have been joyful.

On the day of the wedding, Alexandra and Edith made their way to the downtown apartment of Oncle Edouard. Although my father had originally said he would make do without a dowry, Edouard insisted on paying at least a symbolic amount, to show that Edith came from an honorable family that met its obligations. Edouard would be the one to accompany Edith down the aisle, and he was there, of course, for the signing of the *ketubah*, the religious marriage contract. On that day, as on so many days, he became the father and protector she'd never really had.

There was to be much pageantry and fanfare; close relatives were all assigned important roles. Lily, the fourteen-year-old daughter of Tante Rosée, was the *demoiselle d'honneur*—the maid of honor—given the coveted task of carrying the bridal veil. She was going to walk a few steps behind Edith and gently lift the veil off the floor. Two other young cousins were picked to hold the tall white votive candles, each child flanking the bride as she marched down the aisle.

That was always the most haunting part of the ceremony, the children holding candles that symbolized a long and blessed life.

Edith finally emerged from the room where Tante Rosée was helping her dress. No one was prepared for what they saw—it wasn't simply her beauty brought into high relief by her exquisite gown, which Tante Rosée had sewn by hand.

On dirait un ange, Lily thought, as she practiced holding Mom's veil—It is as if she were an angel.

By the time my mother's family arrived at the Gates of Heaven, most of the guests had already gathered. During weekly Sabbath prayers, men and women sat separately, the men in the large main sanctuary, the women high above in their own separate section on the balcony. But on this day, everyone sat together downstairs, chatting amiably as they waited for the ceremony to begin. The choir—a group of little boys dressed all in white—was singing softly, beckoning the bride to enter God's house. Many of the men, my father included, were in *fracs*—top hats and tails—the women in floor-length gowns. And there was Rabbi Nahum—the most hallowed man in Egypt except for the king—praying silently as he prepared to deliver the benediction.

The newlyweds left for their honeymoon in Ras el-Bar, the resort in the north, where they posed for a portrait shortly after

they arrived—one single black-and-white picture that showed
Edith looking lovely and carefree in a long white kimono and
jaunty beret, while Leon, wearing luxuriant white pajamas, towered over her, his arm draped around her shoulder protectively.

What a perfect couple they seemed—so attractive, so happy.

Alexandra, alone in her apartment on the alleyway, paced
up and down the balcony, puffing on her cigarettes. It
wasn't going well, the marriage. The couple had only been back
from their honeymoon a few weeks and already, my grandmother
had grasped a deep unhappiness from what Edith said and didn't
say in their brief visits together. There was the Syrian mother-in-
law who was invasive and bossy and insisted on speaking only in
Arabic.

Leon was turning out to be demanding and authoritarian and
Edith—meek by nature, *très douce*—was afraid of him. And he was
also set in his ways. He was a man in his forties, after all, used to a
certain kind of lifestyle. Once they had returned from Ras el-Bar
he had simply resumed his old ways, picking up where he left off.

Un mariage de rêves?

Edith was left stranded. There she was without a job, without any means of supporting herself, without even her identity—
Mademoiselle Matalon, teacher, librarian—stuck in that apartment
all day with that old Syrian woman who wouldn't even speak to
her in French. She was utterly dependent on whatever little spending money her husband gave her. Leon would vanish every morning to synagogue, and then to his business affairs, and then leave
again at night.

Alexandra decided to walk over to Tante Rosée's house in
Daher. There, at least she could think clearly, seek advice over
a *café turc*. Her family was terribly upset—the more they heard

these stories about Edith, the angrier they became. When Alexandra briefed Oncle Edouard about the goings-on, he was beside himself. But while both my grandmother and Rosée felt he should have a talk with Leon, admonish him, set him straight, as a man of the world Edouard knew that it was too late.

An annulment was out of the question and divorce was unthinkable. Leon could say that his wife had all she needed—food, clothing, and a fine apartment on Malaka Nazli, one of the grandest street in Cairo.

A month or two into the marriage, Edith informed my grandmother that she was expecting a baby.

4

The Bonesetter
of Mouski

Although Mom tried to protect us from the evil eye, she believed that we were more susceptible to it than other children, certainly when she surveyed the rash of calamities that constantly seemed to befall us, starting with that summer night in 1951 when my brother Isaac got bitten by a rat while the family was away on holiday.

A few years into her marriage, Edith was overwhelmed trying to raise three small children. Her duties as a mother left her no time to dwell on the life she had left behind—her work as a teacher, her bond with the pasha's wife. My sister Suzette, the oldest, was born less than a year after the marriage, a disappointment to Dad who was counting on a son, only a son.

The marriage, rocky from the start, almost didn't recover.

Then two years later, César came and was cherished by my parents as the boy both had wanted and there was a tenuous

Edith with baby Suzette, Cairo, 1946.

peace. Isaac, the baby, was born in 1950. There was a semblance of a family life, especially in the summer.

In those days the family vacationed in Ras el-Bar, a small town in the north, preferring its rustic charms to the urban pleasures of Alexandria. Ras el-Bar was famous for its clean streets and magnificent vistas, especially at the point where the Nile flows into the Mediterranean. Most people rented simple wooden cabanas called *huttes* that were put up at the start of the season and taken down at the end.

Finicky travelers who craved more comforts vacationed in the Aslan, the lone Jewish hotel. It was especially popular for its romantic evening dances on the terrace. Every night, a small orchestra would play waltzes, tangos, and fox-trots, and it felt at times as if all the Jews of Egypt were converging on the dance floor, the men in their white suits, the women in low-cut dresses and the highest heels they could manage.

Mom seemed to fret a bit less on these holidays. Because there wasn't much to do really in Ras el-Bar, my father tended to stay put and not go out night after night as was his wont in Cairo. For the duration of the vacation, he was the exemplary family man. Days were spent at the sandy beach, which my sister loved. Suzette was always upset when, come the late afternoon, the family packed up to go home. "Why can't we stay," she'd wail. It was a time when the family was—in its own way—happy.

And then, the idyll was shattered.

My father, who doted on my two brothers, was in the habit of giving Isaac candy in the evening—*un bonbon*—to help him fall asleep. It did, but late one night, after everyone had gone to bed, the family was woken up by my brother's piercing screams. When Suzette turned on the light, she spotted the large rat and Isaac crying in a pool of his own blood. She, too, started screaming, and everyone realized that Isaac had been bitten by the rat, possibly one fond of Dad's bonbons.

The family packed up and raced back to Cairo, heading straight for the leading medical center, which had a special section for rabies. Kasr-el-Ainy Hospital followed the protocols of the Pasteur Institute in Paris. A century earlier, Louis Pasteur had discovered a vaccine for rabies, and his approach was still the model in this Middle Eastern city where hungry cats and dogs roamed even the most elegant streets, so that the disease— invariably fatal—was widespread. In cases such as my brother's, where you had to assume the animal was rabid, treatment called for a series of painful intramuscular injections administered every day for forty days.

The trips to Kasr-el-Ainy became a family affair. Everyone was traumatized and constantly reliving that awful night in Ras el-Bar—the rat, the blood, the sheer hysteria over what to do.

My sister, who even as a little girl resented Dad, had her own perspective on the sordid affair: She blamed my father and his bonbons. To her mind, the rat had been lured by the piece of candy my brother was holding as he fell asleep. The events of that night in 1951 were played and replayed, analyzed and dissected, and Suzette's disquieting take could never be proven or entirely dismissed.

Mom would declare that the family—and Isaac in particular—had been given the *mauvais oeil,* the evil eye. That was how she understood that terrible night in Ras el-Bar when our entire summer holiday had been wrecked.

The following year, it was my mother's turn to be a victim. Shortly after giving birth to a baby girl we named Alexandra, Mom was diagnosed with typhoid fever. The disease that had killed Indji Cattaui nearly half a century earlier was still the scourge of Egypt. My mom's fever raged and everyone around her gave up hope, especially when the baby caught the disease.

But then a young doctor from L'Hôpital Israélite, the local Jewish hospital, arrived to pull off a miracle. Dr. Baroukh Kodsi was a Karaite Jew who was one of Cairo's leading internists. He was especially adept at treating cases of typhoid and now, unlike the time of the pasha's daughter, there was a powerful weapon at a doctor's disposal—a drug named chloromycetin that was remarkably effective.

Mom survived; my baby sister did not.

It was the Curse of Alexandra, everybody said, being visited on a second generation. It had claimed yet another victim—a newborn infant who'd had the misfortune to be given my grandmother's name and had lived on this earth barely a week.

Bad luck comes in three, my parents believed: *Jamais deux sans trois*—never two without three. Edith, still weak from the typhoid and her loss, wondered what lay in store for all of us, what new calamity would strike the family.

In 1952, seismic changes were taking place all around us, which only intensified Mom's angst.

The year had begun in a terrifying fashion, with mobs setting fires across the city, destroying British Cairo and European Cairo and Jewish Cairo.

The demonstrators had targeted not simply foreign clubs and hotels but Jewish-owned department stores. They had unleashed their rage on Shepheard's, beautiful, elegant, Shepheard's. It burned to the ground, that quintessential symbol of empire—the exquisite terrace, the two bars, the lavish lobby, the tea salon where on that heady afternoon nearly a decade earlier, my father had declared his love for my mother and announced their engagement.

Then, months after the fires, in July 1952, the military staged a coup against Farouk, forcing him to abdicate, and seized control.

To the Egyptians who had chafed under the king and resented the rampant corruption, the revolution was a time of hope and renewal. But to Jewish families like mine, those days and weeks following his overthrow were terrifying—the tanks in the streets, the army everywhere in sight, the nagging, terrifying question: Who will watch over us now?

The king, for all his flaws, had been the Jews' friend and protector, much like his father, Fouad.

Life in Egypt felt very tenuous. At home on Malaka Nazli Street, my parents wondered what to do. Both Mom and Dad had a fiercely Egyptian identity—my father, born in Syria, had lived in Cairo since he was an infant; my mom was a Cairene through and through. Neither could even imagine living elsewhere.

Most devastating for Mom was her younger brother Félix's decision to move to Israel. The two were close—their traumatic upbringing, marked by an absent, derelict father and a mother who was there but not there—had taught them to hold on to each other, to depend on each other. As for Alexandra, who was lonelier than ever in the little room where she resided by herself, she had lived for his visits. What would she do without her charming and unreliable son?

My uncle's mind was made up; as with so many Jews of Egypt in those turbulent years, Félix felt the future was elsewhere.

Still, there were some hopeful signs. General Muhammad Naguib, who was installed as Egypt's president after the revolution, seemed anxious to reassure the Jewish community. The military had a reputation of harboring anti-Semites, but Naguib was mild mannered, courtly, a gentleman as well as a moderate—who felt it was important to cement ties with key constituencies. In a widely publicized visit to the Gates of Heaven, he paid a call on the chief rabbi of Egypt, Rabbi Nahum, and the two men were photographed beaming as if they were the dearest of friends. Then, the general went to L'Hôpital Israélite, another stop in his bid to reassure the Jews of Egypt. As he toured the wards, he chatted with young Dr. Kodsi and other Jewish physicians. Naguib was trying hard to show that the new regime was as loyal to its Jewish population as the monarchy had been.

As events swirled around them, my parents were anxious and distracted, not their usual vigilant selves.

Nobody noticed when Suzette and César retreated to Dad's room one morning to play their favorite game, pretending his bed was a trampoline. It was one of those tall old-fashioned metal beds with springs, and it was possible to jump up and down, bouncing ever closer to the ceiling. My sister would jump, and then César would try to beat her by jumping even higher. The window was open; as they jumped they could see clear across Malaka Nazli Street. They had played this game a million times, and usually my father or mother or one of the maids were somewhere nearby keeping a watchful eye.

Suddenly, without notice, César fell, hard, off the bed and landed on the wooden floor. He began to wail. My mother rushed in and found him all bruised; it was impossible to tell if his arm was broken or badly sprained. My father was summoned.

"Chodi lel barsoumi," Mom cried as our porter rushed to get a taxi—take him to the bonesetter.

This was the way of Old Cairo where there were many medical practitioners without medical degrees, yet who were trusted implicitly. These included the pharmacists, who were learned and could readily dispense advice on common maladies; the midwives, who were taught by their mothers how to deliver a child and were often favored by women over obstetricians; and, of course, the *barsoumis*, the bonesetters. Revered by some as greater than traditional doctors, the *barsoumis* were ordinary Egyptians who practiced a kind of folk medicine. They had no education, no license, yet they were renowned for being skillful and intuitive. Theirs was a secret craft; a boy learned it from his father, who learned it from his father, who learned it from his father. Their knowledge of human anatomy—of the bone structure in particular—had been passed on for generations. Although they lacked many of the tools of modern medicine, they knew how to deal with broken bones as well as any orthopedic surgeon.

Some were true miracle workers; they cared for injured babies and the frail elderly and anyone in between.

César went off with Mom and Dad to the Mouski, the shopping district in Old Cairo where the bonesetter practiced. There were always throngs of people hoping to see him—ordinary *fellaheen* went because he charged very little, but so did more affluent Jews and Coptic Christians, who could have afforded any specialist in the city. But somehow on this day, there were longer lines even than usual that extended outside. My family found the bonesetter in his favorite position—seated on the floor—wearing his usual loose white *ghalabeya* or caftan. Patients waited their turn, and when they were finally face-to-face, they addressed him reverentially as "Ya doctor." Next to him stood his assistant—not a registered nurse but a little boy, probably his own son and apprentice bonesetter. The child would hand the *barsoumi* bandages or pieces of wood that he used as splints. The bandages and wood were the extent of his medical supplies, along with a mysterious cream he'd use as he rubbed and manipulated the afflicted area.

When my brother stepped up to be treated, the bonesetter applied a small amount of the ointment to César's arm. He examined it carefully, delicately, and then shook his head. It was broken in several places, he declared, and there was nothing he could do— my brother had to be taken immediately to the hospital.

My father nodded and paid him his fee—a few coins, because the bonesetter was a humble man who wouldn't dream of taking a lot of money. The *barsoumis* were men of integrity: If a problem was beyond their competence, they knew it and would urge you to go find a traditional physician. But out in the countryside, where there were often no doctors, they were lifesavers. The bonesetters typically attended to large swaths of Egyptian peasants with no access to modern medicine and who couldn't pay for it in any event.

My father put César in another taxi and off they went with Mom to L'Hôpital Israélite, the famed Jewish hospital that was close to our house. In addition to local doctors, the hospital had Coptic Christain doctors as well as many European physicians on staff, including German Jews who had fled Nazi Germany during the war and enjoyed wonderful reputations. But as with any medical establishment, it was the luck of the draw as to who treated you.

The physician who examined my brother was not their most distinguished. He was rash and incompetent—thoroughly unlike the bonesetter who had been so thoughtful and reflective. He

Edith and César as a little boy shortly before his accident, Egypt, circa 1952.

peered at my brother's arm and immediately declared that the only solution was to amputate.

He wanted to surgically cut off the broken arm rather than even try to fix it.

My brother began crying again—he understood perfectly well what the doctor was proposing. My dad, who had a knack for finding top specialists at the drop of a hat, had obtained the name of an orthopedist with an office downtown. My parents grabbed another taxi and, César seated in between them, they made their way to Emad-el-Din, the elegant commercial street near the stock market, where Cairo's leading doctors maintained private practices.

The specialist was reassuring. My brother's case was certainly challenging—he had sustained a complex fracture—but the arm could still be repaired and there was no talk of amputation. He prepared a cast and asked my brother to come in every few days for different procedures. It was a nerve-racking series of encounters, and my mother—who remained oddly calm in these trying circumstances—would stand by as the doctor worked on my brother while Dad sat in the waiting room. When faced with a medical crisis involving their children, my parents tended to get along remarkably well. If any of my siblings were ill, Mom and Dad set aside their differences and came together. Yet in this case, Mom emerged as the stronger of the two. She was no longer the frightened, helpless creature that had defined her persona throughout the marriage, from the day my father had brought her home to Malaka Nazli Street.

It was as if her old toughness—the strength of character she'd shown as a teacher and disciplinarian of the little boys of Cattaui and even the ruffians in Le Sebil—revealed itself in these trying periods. She showed more of what the French call *sang-froid* than my dad, who was fine about paying the bills, but could actually get teary when the going got rough. Edith—though weakened by her own malady and loss—kept her cool in the face of this medical crisis, which included staying with César during every step of his ordeal.

Rabies, broken bones, typhoid fever, the loss of a king, the death of an infant—what next? My mother was on guard for a new catastrophe.

There had been more than three strikes of bad luck, yet she could not rest easy, could never feel the worst was over. We, the children, were the center of her life—why she had stayed in the marriage—but she constantly feared we would be taken away from her. In that sense she was exactly like my grandmother Alexandra: convinced we were in danger and more vulnerable to the evil eye than other children. To her mind, Suzette, César, and Isaac had all the qualities that made them potential victims—they were brainy, beautiful, charming—but instead of simply taking pride in that as other mothers would, she decided that some malevolent force out there was lurking, ready to pounce.

"On nous a donné le mauvais oeil," she would say darkly. Someone has given the family the evil eye. While she was fine during a crisis, afterward she would grow anxious and moody and retreat into a universe of fear and superstition. She, too, was prone to having *des crises de nerfs*—tantrums, nervous outbursts.

It didn't help that she felt profoundly lonely—bereft of friends, lacking any support system except for my grandmother Alexandra, who could hardly give anyone support. With Oncle Félix gone, my mom had lost a companion and sounding board. Her younger brother was flawed, troubled, undependable—but deeply loveable and she missed him terribly. Now that he was hundreds of miles away, living in a country at war with Egypt, it was hard even to exchange letters. Alexandra came over to the house on Malaka Nazli more frequently, careful to time her visits to those periods Dad wasn't around. The two didn't like each other and my grandmother preferred to avoid him altogether. She would venture in timidly, sit on a corner of the sofa, and quietly sip her *café turc*. My mother and grandmother would speak of Félix, how lost they both felt without him.

*M*y father was always out and about—with clients during much of the day, at synagogue in the evening, and then off to his mysterious nightly pleasures, pleasures that didn't include Mom, so she felt completely neglected.

The two never went out as a couple, never saw friends together, never went to a restaurant or to the cinema except with the children.

The truth was that they were so fundamentally different, she had no interest in what he considered pleasurable. Mom could never sit still for a hand of poker, and she preferred to stay home in the evening rather than going out as he did to sample Cairo's ample nocturnal fare. As for Dad's passion for dancing, once upon a time as a young girl my mother had enjoyed the waltz, but she wasn't a dancer in her soul as he was. She was shy and bookish, and terribly retiring.

Her girlhood arrogance, the sense of self that had made her such a striking and memorable figure as she walked the streets of Sakakini, was gone, replaced instead by the melancholy and self-effacement that had overtaken her since the early days of her marriage, when she'd first realized she had made a terrible mistake, and before the children arrived one after the other to keep her busy.

Another woman, in another era, another culture, would have found some means of escape, some refuge, perhaps by confiding in friends, or going back to work. But Mom didn't have any friends. And that golden period when she'd been a teacher seemed so far away. She was only thirty years old, yet she looked back on her years at Cattaui as if they had been the high point of her life, and she assumed that her stint as a career woman was over.

That wasn't merely her depression taking over. A married woman in Egypt could not work. To be sure, even in 1950s Cairo, there were women who held jobs—the pretty salesgirls of Cicurel, the occasional teacher or secretary. But these tended to be young and single or else they were old maids—*vielles filles*—with nobody to support them. For a married woman to seek work was unthinkable, because it meant her husband could no longer take care of his family.

And yet the world was changing. Suddenly, there were women—Jewish women—who were forced to find employment as their husbands grappled with financial ruin. After the revolution, an increasingly capricious military regime confiscated this business or that and asked companies to limit their number of Jewish workers. Men who had been prosperous found themselves penniless and unemployed. But how could their wives even hope to earn a living?

They weren't trained to do much, these Levantine ladies of leisure. Like my grandmother Alexandra, they never anticipated a day when they'd need to support themselves. But several did know how to sew, one of the few practical skills a girl from a good family was taught. The ability to use a sewing machine became a lifeline: In a country where there was little ready-to-wear, women could still find work as seamstresses—*couturières*. And that is how some families with homes and servants—but no income anymore—survived: The wives did piecework.

A couple of years after the revolution, people wanted to believe life was returning to normal, that they could enjoy themselves again.

There were small lulls, periods of apparent stability, then it would seem once again as if Egypt was on the verge of unraveling. A way of life marked by civility and stability became intensely uncivil. There were constant edicts and decrees and policy changes. Some were of epic proportions, such as the decision to redistribute land by limiting the number of acres owned by any one family; others were harsh and cruel—a 1954 law authorized the arrest and detention of anyone who threatened "public order and security," which led to the arrest of leaders of the Jewish community. Others were simply madcap—renaming streets, eliminating all royal titles such as pasha and bey.

By 1954, Naguib—the kinder, gentler face of the revolution, the one who had tried so hard to reassure the Jewish community that all would be well—found himself on the outs. The firebrand colonel who had masterminded the revolution, Gamal Abdel Nasser, emerged as the true power; and he made it clear he had no use for Naguib's conciliatory moves, including toward the Jews. Nasser moved swiftly to unseat Naguib even though the general held the titles of president and prime minister and chairman of the Revolution Command Council.

Lest anyone had any lingering doubt about the brutal nature of the new regime, Nasser placed the fifty-five-year-old general under house arrest.

Naguib—who had helped eliminate the monarchy and its trappings—was confined to a palace that had belonged to a pasha.

People still carried on, went out, left Cairo to enjoy the summer holidays, but it was with a heavy heart, a sense of foreboding.

My family couldn't bear to vacation in Ras el-Bar after the summer of 1951. Instead, we opted for Alexandria, which Dad actually preferred because it was such a lively city with so much for us to do and for him to do. He rented several rooms in a pretty yellow villa at Sporting, one of the most popular beaches in this city of beaches, and where the bulk of Cairo's Jewish community congregated.

Villa Lalouche was owned by a widow, Madame Lalouche, who depended on summer tenants to support herself and her disabled son. The house was old and spacious and elegant in a ramshackle way. It had the advantage of being near the racetrack, my father's great love besides poker. He could go off to the track to bet on the horses during the day and leave my mom and siblings behind to entertain themselves in the relative splendor of Villa Lalouche with its many parlors and lush garden and strange assorted guests, including a beautiful and mysterious

Romanian émigré who vacationed there with her little boy and always insisted on wearing white.

Come late afternoon, my siblings counted on my father to return from the track and take them to any one of the delightful patisseries that were all over Alexandria—Délices, which some said surpassed Groppi's, its archrival in Cairo, and whose windows always had the most magnificent displays of cakes, or popular, smaller establishments such as Athineos or Trianon or Pastroudis that were Greek owned yet specialized in French pastries. Because Dad liked to dine well, he was expansive about going to lovely places with my siblings, treating them to whatever they wanted. There were restaurants all along the Corniche, cafés that overlooked the Mediterranean and served tasty grilled fish and other delicacies. They were invariably thronged.

Sporting Beach was only a block or two away from the Villa Lalouche, and Suzette was thrilled—as long as we were near the sea, she was happy. Back at the villa, she and my brother loved to go down to the garden. It had a jasmine bush, and my sister could spend hours gathering jasmine and then making necklaces out of the flowers. A few feet away, there was a small guava tree César favored. He would reach for a guava and eat it there in the garden, and nothing ever tasted as delicious as the fruit he had picked himself.

My sister was also reunited with her friends from school and would join them in their fancier enclaves. Alexandria was divided into a series of beaches, and some were snazzier than others. Sporting was great fun and lovely, but the cool crowd gathered at San Stefano and Sidi Bishr and whenever Suzette had an invitation, she'd find a way to go.

Even as my sister tried to flee—to the beach, to be with other families, to the garden—she was summoned back to deal with the dramas at home.

Dramas that now flared up even on vacation.

My mother had sunk into a hopeless mode, and every few days she had another outburst—*une crise de nerfs*. Her usual coping mechanisms—the edifice of rituals and superstitions that she had built to protect herself, to give herself some measure of stability— were crumbling. The prayers and fasts and incense burning she relied on to chase away her demons and ward off the evil eye were no longer enough. She was spiraling out of control—feeling abandoned by her husband on the one hand and, on the other, overwhelmed by three young demanding children and the searing memory of the death of a fourth.

The last couple of years had taken their toll, I suppose. It was hard to predict what would ignite her rages and bouts of hysteria, but they were getting more frequent and more frightening. It was odd because by day, she could be the most dutiful of wives and mothers, rounding up the children to go to the beach, making sure everyone had enough to eat "et plein de distractions"—lots of fun diversions. But then they'd come home, and the darkness would set in even in the light-filled Villa Lalouche.

César and Suzette were the main witnesses to her boundless grief; Isaac was too young to understand, and my father was usually out or would leave at the slightest hint of a scene. Mom could work herself up into a frenzy, begin to scream, pull at her hair, slap herself again and again, her hand beating her cheek in wild, hysterical motions.

My siblings would watch petrified, then one or the other would take action. César would run and get cold water and throw some on her face simply to shock her into calming down. Suzette tried to soothe and reassure her simply by talking to her.

Mom's pain reached a crescendo that summer at the villa, and one day when she was especially distraught, she threatened to kill herself.

No one knew exactly what had unhinged her. Had Dad stayed out too late? Had the children been unruly?

My sister watched horrified as Mom grabbed a bottle of tincture of iodine that she used to disinfect our cuts and bruises and lifted it to her lips, as if she were going to drink it. Suzette rushed over to stop her. But even as my mother seemed to be going over the edge, she retreated, and it was never clear whether the scene with the bottle of *tincture d'iode* was a sign of the depths of her sadness or a cry for help or merely a show of histrionics.

Later my mother appeared subdued but fine. She said only that someone must have given her the *mauvais oeil*, the evil eye.

That was the only explanation tendered for her despair. There was really no one to help a depressed housewife except her immediate family. In this case it meant that my siblings found themselves playing the same watchful role my mother had as a little girl when Alexandra was falling apart.

Even as young children, they became adept at calming her and bringing her around; what choice did they have? In this culture where rabbis served as marriage counselors and there were no psychologists or social workers, there were few options; anyone too far gone was sent to the Yellow Palace, the insane asylum. There wasn't even a medicine man like the bonesetter of Mouski who could have helped Mom, who could have applied a soothing cream to her fragile soul and set her straight and made her strong again.

5

The Porcelain Dolls
of Malaka Nazli Street

On the morning of November 14, 1956, Dr. Baroukh Kodsi reported to work as usual at L'Hôpital Israélite. Cairo's only Jewish hospital, which was located by Malaka Nazli Street, not too far from our house, was filled with patients that day. Many of them were quite poor. They crowded the wards of the *troisième classe*—the "third class" as the indigent service was called, as if the medical center were some great ocean liner offering different cabin sizes and meals depending on your budget and social rank. The tall, soft-spoken internist who had pulled Mom from the brink a couple of years earlier when she'd battled typhoid fever was much in demand, rushing from patient to patient.

He was in the midst of doing rounds when the news reached him. The Egyptian army had taken over the hospital. Soldiers had encircled the building and were about to enter. The doctors—

nearly all of them Jewish—were ordered to leave the premises immediately.

Some of the nurses started crying. Patients looked on bewildered from their beds.

All the physicians lined up and dutifully made their way to the door. They were warned to take nothing with them except their stethoscopes—not their patient charts, not their little black bags or their medicines or syringes. As Dr. Kodsi walked out, a soldier ordered him to raise his arms in the air, peered at the silver stethoscope he was carrying around his neck, and waved him out the door.

Then it was over. The Jews of Egypt were no longer in control of their own hospital. Dr. Kodsi was told he could not reenter the premises again. *What will happen to my patients?* he wondered as he stepped outside.

The fall of 1956 was a season of terrible endings and one small beginning—mine. I was born in September; one month later Egypt was engulfed in a war with Israel, France, and Britain over control of the Suez Canal.

The Jewish community was on tenterhooks. How were they going to cope in an increasingly hostile atmosphere—a country where it was clear they were no longer welcome? Many didn't wait to find out. Within weeks of Suez, they started leaving in droves— faster even than after Farouk's overthrow four years earlier when the siren song of the revolutionary council and General Naguib kept people vaguely calm and hopeful that the charmed life they had known all these years in the Levant would somehow continue.

But Naguib was languishing under house arrest, and as Nasser consolidated his power and deepened his bonds with the Soviets and Eastern Europeans, any illusion that all would be well disappeared.

It was all a matter of how long you could hold out. Those families who held European passports left immediately: They had no choice and were given a couple of weeks and at times a couple of days to get out. Others lost businesses, jobs, livelihoods, and they, too, were forced to flee.

Dr. Kodsi found that after the expulsion from L'Hôpital Israélite his practice dwindled. Most of his wealthy Jewish patients had left, and his Muslim and Christian patients were avoiding him now. Going to a Jewish doctor was suddenly taboo in this society where once upon a time, it had been perfectly natural for a Muslim to see a Jewish physician, or a Jew to consult a Coptic Christian specialist, while followers of all faiths sought help from their local *barsoumi*—the bonesetters who were native Egyptian Muslims.

Families who had never ventured far beyond Sakakini or Suleiman Pasha found themselves headed to the ends of the earth. When Australia let it be known it would take in Egyptian Jews, lines formed day and night around the Australian Embassy in Cairo as Jews anxious to secure visas for themselves and their loved ones vied for immigration papers. Then came the rumor that Brazil would also welcome Egyptian-Jewish refugees, and there was a mad rush to its mission as talk in the community focused on what life would be like in Rio and São Paulo.

Cairo turned into one big yard sale as families placed their favorite belongings in the street and invited strangers to come purchase their precious furnishings at bargain basement prices or simply gave them away.

Dr. Kodsi and his fellow doctors decided not to stick around. Offered positions in America, they fled, and the community was left with only a handful of physicians and specialists to take care of emergencies.

It was into this chaos and confusion that I came along. Our friends and relatives were also leaving, but it was as if I were holding my family hostage. Impossible, my father would say ruefully when asked when we were planning to depart: I was too small to travel, too fragile. My mother echoed him and began referring to me as *pauvre Loulou*, poor Loulou, and the nickname stuck.

Our relatives left one by one. These were wrenching good-byes because no one really wanted to go and no one felt they had any choice. One after another would make it safely to Israel—the favored destination—and we'd feel so relieved. But then, within a few weeks or months, the sad news would reach us: this uncle had suffered a heart attack and died; that aunt had succumbed to a vicious cancer. And, no, of course, these disastrous, unhappy endings couldn't all be blamed on their displacement; and yet that is what we felt in our innermost beings—that leaving Egypt and finding themselves stranded in a strange land had been ruinous for our loved ones.

My grandmother Alexandra mournfully took off for Israel to join Oncle Félix. She had never been separated from Mom for more than a few days. How would she possibly survive? Even now that Edith was a grown woman, when she walked with my grandmother through the streets of Cairo, she still held Alexandra's arm tightly, as she had as a young girl. Alexandra was a regular visitor to our house, and she would happily have lived with us if Dad would only have allowed it.

Away from Edith, my grandmother was barely coping. In the veiled communications that reached us—we weren't allowed to get direct letters from her since Egypt and Israel were technically at war—we gathered that she was desperately unhappy and longed to be reunited with us again.

We concluded that no matter how dangerous and unpleasant Cairo had become, the world beyond it was much worse. And so we lingered, hoping against hope that the turmoil brought about by Nasser and the revolution would pass.

In those early years, I was often in the care of my father, a won-

*Alexandra around the time she left
Egypt and all that she loved to emigrate
to Israel, circa 1957.*

drous babysitter even after suffering a terrible accident. Dad had
fallen in the street a year or two after I was born, and his hip was
shattered. Because he was so tall, his injuries were particularly
severe.

Of course, L'Hôpital Israélite was no longer an option. The
military controlled it now, and all the Jewish patients had been
transferred to other facilities. Dad landed in a small careworn
public hospital known as the Demerdash. He spent months and
months there, often in a state of excruciating pain. After being
discharged, he was essentially homebound, confined to his room
overlooking Malaka Nazli Street.

Though he was in constant pain, he could still keep an eye on
me, entertain me, share with me the little snacks he ate continu-
ously. He relished the fresh figs and tangerines he bought from the
vendors who came directly to our balcony, or plates of feta cheese
and olives, along with his favorite, cans of sardines steeped in olive
oil that he feasted on every day.

Pouspous, our cat, hovered nearby, and my father tended to

include her in all our activities. When we ate lunch, she ate lunch, settling on his knee and delicately reaching with her paw to the table for whatever goody he offered her.

When we sat on the balcony, she made herself at home in his arms or mine. Dad taught me the simple pleasures of taking in the street life in all its splendor. Mom had no patience for that but I thoroughly enjoyed it, and Pouspous also seemed riveted by the rushing crowds and the whirl of traffic. Strangers liked to stop and speak with us because Cairo was a friendly effusive city, and I could have sworn Pouspous was following the conversation and would have joined in with a choice comment or two if only she could.

As the oldest sibling, Suzette was a fallback babysitter. Her idea of taking care of me was simply to take me along whenever she went off to see her friends. Fleeing the turbulence at home, the tensions that still flared between my parents even now that

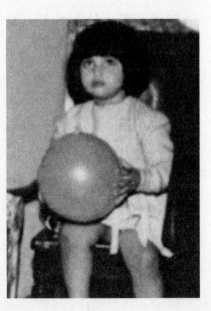

Loulou as a toddler at a party, Cairo.

Dad was incapacitated, my sister found refuge with the Wahbas, a family of three sisters—one prettier than the next—and two brothers. Their parents emblemized the ideal of a happy marriage and a proper home life. The Wahbas lived not far from us on Malaka Nazli though they were already looking to move *en ville*—downtown—where the fashionable set lived. It was a defiant gesture, of course, because Jews were leaving, not finding themselves nicer homes. The Wahbas embraced my sister, showered her with love and understanding, and gave her the peace of mind she lacked at home with us.

The Wahbas had a nickname for me: *la petite poupée en porcelaine,* they called me, the little porcelain doll. Suzette had a different take; she dubbed me "The Assassin," because of a particularly dark look I would get when left on my own. It wasn't at all that of a doll—I was an anxious, resentful child, both fearsome and fearful at the same time.

Mom wasn't so much a part of my world those early years. I'd constantly see her running out of the house on her mysterious errands, a whirling dervish of activity, seemingly incapable of stopping. And even when she was back at home, usually by the late afternoon, she would be tutoring Maggy Wahba. Maggy was a whiz at math but struggled with her literature and composition classes. She'd come after school and settle by the kitchen, and as my mother cooked, Maggy would wail about her latest bewildering assignment from the Lycée Français. Mom would buck her up, remind her of the wonderful grades she received in all her other subjects, and then, as she prepared dinner, she'd dictate the requisite essay. That is how she kept her hand in teaching so many years after she had left L'École Cattaui.

There were four children now, each one of us demanding in our own way. Suzette as a teenager was always angry and clashed constantly with Dad—both were very stubborn. Suzette found my

Loulou being held by Suzette, with Edith standing nearby, at a children's party thrown by her cousins, Cairo.

father far too authoritarian, and his efforts to control her invariably backfired: My sister only became more alienated, rejecting whatever he valued, especially religion. Mom was caught in the middle, unable to placate either. My sister was her biggest challenge, always seething and rebellious and on the verge of running away, except that there were no runaways in Cairo, so all Suzette could manage were her escapes to the Wahbas and her other constellation of friends. César was more docile, more malleable, and Mom tended to view him as a godsend from the start, the honest broker who even as a child seemed capable of getting along with everyone. Isaac, like my older sister, was moody from the start.

Then I came along and Mom, who had always bitterly resented the endless dreary household duties brought on by motherhood, found that she could pawn me off to my father or sister and go on about her business, though I was never exactly sure what that business was.

César turned thirteen a year or so after Dad had his accident, and his bar mitzvah, or coming of age, was a subdued affair. Unlike his friends, he didn't have a party—only a small gathering at our house. But at least the ceremony was at Temple Hanan, a grand synagogue near our house that had tall vaulted ceilings and a whisper of the opulence of old.

My oldest brother had changed forever after Dad's accident—our entire family had. He found himself constantly looking out the window on Malaka Nazli Street, anxiously waiting for one parent or the other to come home, wondering if some catastrophe would befall them.

When I started school in 1961, my mother took charge again. She decided to enroll me at the Lycée Français de Bab-el-Louk. The lycée had a vast campus with a private courtyard. The school was sealed from the outside world by a big gate that was kept shut except when we arrived in the morning or when it was time to leave in the afternoon. Children would gather in the yard and one by one, their parents, maids, *bawabs* (doormen), or chauffeurs would swoop down to take them home.

César or Suzette usually took me to school in the morning, but Mom was expected to come pick me up. Yet day after day, I found myself the last child at the lycée, anxiously waiting for her to appear.

As I stood alone in the vast courtyard, my imagination ran amok. Had my family forgotten all about me? I was sure that was it—that in the midst of the hullabaloo about staying or leaving Egypt, my parents had lost track of me. What would happen if no one came? I was scared out of my wits. It is not as if the lycée staff was especially solicitous, and I suppose the administrators thought we were perfectly fine on our own in the courtyard. Occasionally, some staff member would come down and notice me waiting, anxious and teary-eyed. Most often I waited alone.

At last, even as I'd given up hope, I would see my mother arrive, a small figure pushing past a massive wooden door. Smiling and vaguely apologetic, she'd have in her hand a little bag of *petits pains,* the soft, warm golden miniature rolls she had stopped to purchase at a bakery on her way to pick me up. As we walked together to the tram, she'd hand me one of the rolls, no bigger than my fist, and they tasted utterly delicious after the long school day and the rigorous classes and the agony of not knowing if anyone was ever coming for me and my terror that I would be left to fend for myself forever at the Lycée Français de Bab-el-Louk.

This scene repeated itself day after day, week after week. Mom was habitually late coming to get me and I was always so nervous, practically on the verge of hysteria, convinced that my own family had abandoned me in their distress over the revolution and its aftermath. She never explained why she was so late, and I was never sure she or anyone in the family was even aware of the intensity of my distress.

The political situation only exacerbated Mom's inability to cope. It was such a turbulent, confusing period. We weren't leaving and we weren't staying and we didn't know what we were doing. My family was so preoccupied that my childish concerns were by necessity given low priority.

The ultimate snafu came at the end of the year. Each June, the school held *la distribution des prix*—the ceremony when prizes for academic performance were given out to top students. On that day, I was home with the maid: My family had completely forgotten about *la distribution des prix.* After being contacted by the lycée the next day, one of my siblings went to pick up my prizes—a couple of children's books, beautifully wrapped and adorned with ribbons—and brought them home to me.

I think that my mother felt terrible—how could they have allowed me to miss that special day? She tried to make it up to me with a trip to our favorite dressmaker. Even as a young child, I had an immense passion for clothes, and trips to *la couturière* were obligatory for any Cairo female of a certain wealth and social status. Everyone had their personal dressmaker who could whip up the latest Parisian designs, since this fashion-conscious city fancied itself as European even more than it was Arab.

We favored a trusty middle-aged woman with her own small workshop steps away from Sakakini Palace. As a rule, I was blissfully happy at her *atelier*, a universe of fitting rooms and sumptuous fabrics and women delicately wielding tape measures.

But on this trip, because I had developed a fixation on a black dress with a rose at the waist, all hell broke loose.

It began one night at my father's synagogue, which held a get-together to restore morale among our dwindling community. We jammed into the little shul to watch a film of a fashion show in Europe. I was riveted by the spectacle of tall elegant models strutting up and down a runway, clothed in seductive dresses with low backs and plunging necklines. As I looked around me, I noticed the women had dressed up for the occasion—as if they were competing with the models on the screen. Several wore clingy black dresses with a large satin rose pinned to their waists.

These struck me as the epitome of style—so grown-up and confident—and I longed to look exactly like them. When Mom took me to *la couturière* and I was asked what I had in mind, I declared that I wanted a fitted black dress with a rose at the waist.

It was completely inappropriate, of course; even in modern Western cultures, little girls didn't wear black, but that was especially the case in Egypt, where young or old, women traditionally avoided what was seen as the color of sadness, the color of mourning.

My superstitious mother took pains to remind me, as she had a hundred thousand times, that I shouldn't even want to wear black. "Loulou, ça va porter malheur"—it will bring you bad luck, she kept saying.

The dressmaker offered me a choice from a selection of outfits she had recently made for other children. She held up a darling hot pink dress with a low waist that was clearly a popular model— and which I secretly found very pretty indeed. But I declared I would have none of it—I didn't want to dress like other little girls, I fumed. She turned to my mother with alarm as other clients watched aghast.

I think that everyone was taken aback by my fury. We left with nothing resolved, except that I felt more anguished than ever.

When we returned a week later, the dressmaker anxiously brought out a pretty lime green dress she'd sewn specially for me. She had made it with one of the fabrics my mother had picked out when I wasn't looking, and, in an attempt to be conciliatory, she had put a small rose at the waist. I peered at it suspiciously, then tried it on; it fit perfectly, and I agreed it looked quite wonderful and my dreams of owning a sleek womanly black dress were temporarily forgotten.

Unfortunately, at around this time, I became very sick and no one knew why or what to do about it, though Mom was certain it was bad luck. She had warned me, after all. The doctors we consulted seemed completely at a loss. They peered at the suspicious lump at the top of my left thigh and then at me and shook their heads.

Despairing, Edith took me one day on a long journey to Fustat, the most ancient part of Old Cairo. We were headed for Ben Ezra, the synagogue that was hundreds of years old; the point of going there wasn't really to pray but to mingle with the poor who gathered there and give them alms. It was considered a blessing—a sacred, hallowed deed—to give them charity. The tradition of helping the poor of Ben Ezra dated back centuries, and my parents were convinced the mendicants had a direct pipeline to God. My father had taken Suzette with him when she was a little girl, giving her coins or small bills to offer the people who showed up near the synagogue and seemed so pitifully in need. Now, Mom was urging me to stuff lots of coins directly into the hands of the people who shyly hovered around us, and in turn they vowed to pray for me and God would listen to them.

Dad shared her belief in the mystical powers of the poor of Ben Ezra. At least once a month, my father took his clothes—his beautiful white sharkskin suits that had become slightly frayed, shirts, trousers, vests—and gave them to the men hoping for a handout, and who were delighted to find they not only had some coins to get by but a resplendent new wardrobe.

Mom and I entered the virtually empty synagogue and recited a small prayer. It was always dark inside Ben Ezra—the building itself was ancient, dating back to a time when Jews lived in this decaying part of the city and, legend had it, to the spot where the pharaoh's daughter had found baby Moses floating in a basket.

On our way home, we hurried past a beggar sitting on the sidewalk with his two stumps of legs exposed—they were completely bruised and covered with scabs so that his limbs even from a distance were a bloodied red and seemed to glow in the evening light. I was hypnotized and kept staring at the poor man's hopelessly damaged legs, even as Mom tugged at my arm and tried to rush me along streets of a Cairo I didn't even recognize, a Cairo that I found strange and frightening.

I kept looking back at the beggar. I had the feeling that in this part of the city, he was not an uncommon sight, that if we kept walking I'd see other beggars in his condition, like the thin, ailing cats that seemed to be everywhere and that looked nothing like my well-fed gourmet of a tabby, Pouspous.

My mysterious ailment, which we knew only as cat scratch fever, flared up several times in the months that followed, and then seemed to subside once and for all, perhaps because of our pil-grimages to Ben Ezra and other Jewish shrines. By the time my youngest brother Isaac turned thirteen, old enough to have a bar mitzvah, we knew that our most fervent prayer would go unan-swered.

It was the spring of 1963, and we had worn out our welcome in Egypt. In the eyes of the new regime, we weren't Egyptian and never had been, even though we were born here, even though Dad had lived here since arriving as an infant from Aleppo in 1901. We would have to leave.

Our visits to *la couturière* intensified. It wasn't about ordering beautiful dresses anymore: She was sewing far more prosaic items for me, flannel T-shirts by the dozens, and woolen slacks such as I had never worn. A colleague of hers was commandeered to knit me

On the eve of leaving Egypt, Leon and Edith had a family portrait taken with Isaac and six-year-old Loulou.

some sweaters—thick pullovers that were several sizes too large. Wherever we were going, it was going to be nothing like Egypt.

The fabric of choice for my new wardrobe was *du castor*—a kind of flannel typically used for pajamas or robes, and that was about as heavy as fabrics got in this sultry clime we took for granted. My mother was persuaded that I had to have as many models of *des robes en castor*—flannel dresses—as our seamstress could whip up.

Overnight, I found myself with a collection of navy blue crew necks and flannel dresses. I wondered why my clothes were changing so dramatically overnight. This is how I learned that we were leaving Cairo, how I began to grasp the intricate mechanics of exile: through my awkward, bulky *robes en castor*.

Those last few days, all of us—with the possible exception of Pouspous—were nervous and on edge. One morning, my father went out with César, and the two came home brandishing an expensive transistor radio, a Sony, sleek in its brown leather carrying case. Wherever fate landed us, we were at least going to be well informed and understand the world beyond Egypt. My father was clearly despondent. He couldn't bear to leave, couldn't even imagine how he would cope.

César took possession of the radio and carried it around the way a woman holds on to a valuable shoulder bag. It gave him a sense of safety, I suppose: Those days my oldest brother was gripped by fear—fear that we were leaving, fear that we would not be allowed to leave.

Unlike the rest of us, a strange calm descended on my mother. She loved Cairo every bit as much as Dad, but the city of her youth hadn't been the same since Alexandra and Félix left. There had been so little happiness in the years since my grandmother had joined my uncle.

At last—if only we made the decision to settle in Israel—she would be with them once again.

Edith with official papers allowing her to stay in Paris for only a few months, 1963.

Of course, we had made no such decision. And Mom was too meek and too passive to insist on it no matter the hardships. We were leaving Cairo without the slightest idea where we were settling permanently. We would first journey to Paris, the most popular transit point for refugees headed for God knows where.

One afternoon before we left, Mom took a walk to Sakakini. She hurried past the palace and made her way to the Alley of the Pretty One and the corner house where she had grown up. Its newest occupants were seated on the terrace exactly as she and Alexandra had done so many times, and she felt a strange comfort knowing that life hadn't really changed in this corner of Cairo, that revolution or not, the favored ritual of her youth, gathering for *café turc*, was popular with a new generation. As my mother watched them sipping their small cups of coffee, the sadness passed, replaced with hopefulness: She would see Alexandra again, enjoy a coffee with her again on some other terrace in Tel Aviv or Haifa, hold on to her arm again.

That was all she needed to believe to weather the storms raging around her—my father's anguish at leaving, César's terror, Suzette's balkiness, Isaac's moody blues, and my lingering, mystifying illness. All would be well once she was reunited with her mother.

We left Egypt on March 17, 1963, and made our way to France. We had barely settled in our dingy refugee hotel when news reached us that Alexandra had died. Though we had no money— we had been allowed to take only about two hundred dollars out of Egypt—Mom purchased a black dress, simple and very inexpensive. She wore it every day, mourning Alexandra in the only way she could in this new friendless city that didn't really want us and where we didn't belong.

On many days we huddled around the radio. It seemed to be the only link between us and the world. We followed the news with an odd sense of detachment, a kind of numbness. Edith Piaf died, and hundreds of thousands of people turned out into the streets of Paris to mourn her. We stayed put in our hotel room. President Kennedy was assassinated, and again, we remained holed up in the hotel, debating where to settle, Israel or America, America or Israel, or—as I kept urging—simply returning to Cairo.

Mom emerged briefly from her mourning to compose one of her startlingly beautiful notes to French officials. She hoped to persuade them to let us settle in France where at least we knew the language. She had no desire to go to America, and at least by staying put here in Paris, she thought she had a fighting chance of seeing Félix again. Her brother had reinvented himself as a journalist in Israel and occasionally traveled to Europe on assignments.

By December it was clear we weren't going to be allowed to

stay in France. Dad, after heavy lobbying from my oldest brother, decided we weren't emigrating to Israel, either.

For my mother, the last flicker of hope was lost.

We finally heard from immigration authorities that we would be resettled in America. My father signed papers promising to pay back the cost of five and a half third-class tickets aboard the *Queen Mary*, bound for New York.

BOOK TWO

*Rebuilding
the Hearth*

BROOKLYN: 1964–1970

6

The Legend of
Agent Extraordinary

From my earliest days at the Shield of Young David, I relished my role as resident contrarian. I used the women's section as my perch to challenge and question and dispute many of the notions those around me held dear.

Nothing was more sacred to our congregation than the need for a woman to marry and marry young. No matter that the rest of America was moving in a different direction and beginning to challenge the need for a woman to get married at all.

Here, an engagement while still in high school was the norm; that way, the wedding could take place immediately after graduation.

There were any number of eligible young women who sat with us behind the divider. Their quest for a suitable mate was of intense interest to all of us, a subject of much discussion, not to mention concrete matchmaking efforts on the part of the older matrons. Some saw it as almost a religious obligation to marry off any single female.

At sixteen, Marlene, a pretty brunette with a sleek page-boy and an elegant wardrobe, was my mother's favorite in the women's section. Mom liked Marlene for her gentle demeanor, though she was actually quite forceful and outspoken. Ever since enrolling at the House of Jacob Academy, a religious girls' school in nearby Borough Park, known as the strictest *yeshiva* in all of Brooklyn, Marlene had become increasingly devout. She changed before our eyes, arriving each week in dainty skirts that fell below her knee and modest—though stylish—sweaters. If she were husband hunting, or was even contemplating marriage, you couldn't tell from her sedate behavior. She came early, prayed intensely, and seemed to prefer our company to that of the men across the divider. Yet we knew that Marlene would need to be engaged within a year or two because rules were rules at the Shield of Young David.

Fortuna, Gladys's younger sister, couldn't have been more different. Pretty and, to my eyes, a tad arrogant, she ensconced herself in a prominent seat that offered a good view both of the men's and women's sections. She wore heavy eye shadow, and her black hair was teased, sprayed, and pouffed provocatively. Her skirts were short and her sweaters fashionably clingy.

Yet I realized that she was, in her own way, a hopeless romantic. Even as she dreamed of that elusive mate, she pined after one man who was far from the Shield of Young David and wasn't even Jewish. Fortuna had a crush on Elvis Presley. She went to all his movies, often in the company of Gladys, bought his latest records, and memorized the lyrics to his songs.

I noticed that she became animated whenever she spoke about him. I suspected that if Elvis were ever to materialize in Bensonhurst, Fortuna would happily abandon her family, her watchful brothers, her doleful sister, and the women's section to run away with him to Memphis.

Another one of the eligible young women behind the divider was Janet, who was brainy and attractive. Janet had had an unfortunate marriage that had lasted only one day. We never found out what happened, but it was the source of gossip, the kind of experience that could ruin a girl in our community. But Janet was charming and educated, and her mother was one of the saints of our women's section, the kind of person who visited you if you were sick and made a match for you if you were single. Janet waited for a new opportunity.

I had very different ideas about the need for women to go husband hunting when they were still so young.

"I don't plan to marry till I am at least thirty," I airily announced one Saturday afternoon in the women's section to no one in particular.

"Then you will die an old maid," Estrella, my friend Gracie's aunt, shot back.

Edgy, fun-loving Estrella was from Mexico, where a large contingent of Syrian Jews had settled, and she had an acid tongue.

"Thirty is a woman's most beautiful age," I replied, echoing a line that my older sister, still unmarried, had been spouting of late.

"See who will marry you when you are thirty," Estrella sniped as the rest of the women burst out laughing. Although their reaction stung, they weren't being hostile or malicious. On the contrary, most were sweet, traditional women, waging a hopeless battle to hold on to the ways of their old cultures in the wilderness of New York.

A hauntingly pretty woman with large dark almond-shaped eyes and high cheekbones turned around and told me quietly, matter-of-factly, as if she were speaking to another adult: "I don't think you will feel that way in a few years. I think that you're wrong."

I was startled into silence. The woman we knew only as the Great Beauty wasn't a regular in our congregation, yet here she was chiming in to the group discussion. She made her point thoughtfully, while examining me closely, as if trying to figure out when my pretty little-girl face would grasp the enormity of growing older. I was so surprised she was even talking to me that I merely nodded meekly, my usual bravado gone.

I experienced Mrs. Menachem very differently; whatever she said made me bristle.

I am not sure exactly when it all began, but from my earliest days at the Shield of Young David, Mrs. Menachem seemed to take a dislike to me. She would make me the target of her taunts and admonitions. Originally from Israel, she was a devout woman who covered her head, which was more the tradition of Eastern European Jews. She favored stiff wigs that made her stand out from the other women who wore at most a flimsy scarf or small lace kerchief on their hair, in the tradition of the Middle East where the pursuit of modesty was a theoretical exercise.

Oddly, she and my mom enjoyed a perfectly amicable relationship. They rather liked each other. Mom viewed her as a formidable woman who was trying to anchor us in this post-Cairo exile of ours where there were few standards and we all felt unmoored. And I am sure that Mrs. Menachem felt sorry for my mother and saw her as sweet and refined, overwhelmed by having to raise a difficult daughter in rough, uncertain terrain.

Mrs. Menachem launched a battle to persuade Mom and the other mothers to enroll us at the House of Jacob Academy. Madame Marie, desperate to find a way to tame her rebellious daughter, dispatched Celia in a uniform to Borough Park to mend her ways. In my case, Mrs. Menachem made an even more impassioned plea. Only at Beis Yaacov—the House of Jacob—she told my mother, with its single-minded focus on discipline and faith, did I have a chance of escaping the path I seemed so intent on pursuing, the path of an assimilated young woman growing up in an America even Americans found troubling and bewildering.

Assimilation—such a natural process for most new immigrants seeking to make their way—was anathema to my community, to be avoided at all costs. Mrs. Menachem, the schools she advocated, and, of course, my synagogue were all trying to create safe harbors, to keep us shielded from the temptations of this alluring culture beyond.

My mother was sorely tempted. Mrs. Menachem arranged an interview with the principal. He looked me up and down, clearly disapproving of my casual appearance in a short blue-and-white knit dress, then sighed and agreed to admit me: I was so clearly a lost soul and he had been briefed about my dire situation by Mrs. Menachem.

There was a catch. The students at the House of Jacob studied Hebrew as intensively as English. Since I only had a slight knowledge of the language by their standards, I would have to take Hebrew classes with the kindergarten students.

I balked to no avail. Off I went to the House of Jacob to attend classes with pupils several years my junior.

I lasted five days. By then my sister had staged an intervention. She warned Mom that if she let me continue at the House of Jacob, I would be married and wearing a wig by the time I turned eighteen. My mother found that sobering. She had no desire to see me rushed to the altar at an early age.

My father stayed out of *L'Affaire de la Yeshiva*. I am sure he would have loved to see me enrolled in a Jewish parochial school and immersed in religious studies. But he didn't weigh in very much on my upbringing anymore. Since we'd arrived in America he had suffered a drastic loss of confidence. He had been a prosperous, aggressive businessman in Cairo, a lover of fine restaurants and white sharkskin suits. But here in New York, unable to find meaningful work, he had become a tie salesman, wandering the streets and subway stations selling imitation silk ties. He barely scratched out a living, and in his humbled state, he became, as it were, mute. He spent more and more time at the synagogue down the block, the Congregation of Love and Friendship. He deferred to Mom on most questions about my upbringing, including my education.

The following week I was back at PS 205, my public school around the corner.

From then on, Mrs. Menachem regarded me as a lost cause and, worse, dangerous. She was determined to stop me from spreading my subversive ideas to other girls and began to quietly approach mothers of my friends.

She'd contact them during the week to urge them to keep their children away from me. I was shocked and distraught when I learned of the phone calls. I realized that if she could, Mrs. Menachem would destroy all my friendships.

But the women behind the divider were a loving bunch, each one a surrogate mother. They shook their heads at my antics and seemed mostly amused by my epic battles with Mrs. Menachem. I offered a welcome diversion from Rabbi Ruben's solemn speeches and the honeyed chants of the cantor.

They also liked me. In a congregation of foreigners, she was the foreign one, the woman who insisted on covering her hair. They simply ignored her warnings and let their daughters play with me, and I was more popular than ever.

I stuck close to Gladys, who loved to cook, to eat, and to feed me. I had noticed her crying during the services—we all did—but no one, not even Mom, was willing to explain to me why she was so often distraught. Finally I persuaded one of the women behind the divider to enlighten me.

Gladys, married to a quiet man named Saul who was some

years older than her, hadn't been able to have a child. She was completely grief-stricken. If she could produce such magnificent kiddush meals, enough to feed an entire congregation, why couldn't she produce a small infant? Unable to find a satisfactory answer, she wept, prayed passionately for a miracle, and cooked for us abundantly.

We learned to expect the most grandiose meals precisely when Gladys seemed the saddest and had no recourse but to flee sobbing into the kitchen. I suppose that she longed so much for a little girl of her own that she adopted me and tried to shield me from Mrs. Menachem, and to fatten me up with bowls of all-American tuna fish.

I had been in love with Maurice since I'd first arrived at the Shield of Young David and glimpsed him beyond the divider. I lived for Saturday morning when I would enter the women's section and find him standing there, quietly praying in his corner on the other side of the wooden barrier. He was to me the epitome of style—manhood in a dark red blazer—though he never spoke to me, never so much as acknowledged me.

The rest of my week—school, teachers, my classes, my friends—seemed vague and unimportant in comparison.

Nothing mattered as much as that moment when I would be reunited with him, with only that slender partition between us.

The intensity of my feelings, the fact that I could have a full-blown crush on this older boy and think of little else but him, would surely have taken aback the adults around me had I ever confided the full extent of my yearning. They would have been troubled by the unremitting nature of my attraction, the fact that I'd entertained a rich fantasy life and had spun a romantic web around this boy since I was eight years old.

Maurice was Marlene's younger brother, and the older brother of my closest friend behind the divider, Diana. I didn't breathe a word about my feelings to either, at least not at first. In Marlene's case, I suspected that she and the other grown-ups who were always hovering near us would be dismissive, that they'd laugh at me like Estrella, and see this as another fanciful ploy of mine to get attention.

Could a little girl fall in love? Was a child capable of great passion?

Hard as I tried, I could never get Maurice to notice me. Perhaps the wooden separation really worked: I didn't receive more than a brief, absentminded nod—if that—a fleeting recognition that I was there Saturday after Saturday.

Maurice had a close friend, Joseph Hannon, who was also handsome and reserved. The two boys sat near each other, and I could see them occasionally speaking quietly together. How I envied their bond. When I went to his sisters' house Saturday afternoons, I would find the two of them reunited, chatting or shooting pool at the small pool table Maurice had purchased. I longed to be asked to play, or at least join in their conversations. But Maurice, whether at synagogue or in his own home, seemed to inhabit his own world, a world that excluded pretty much everyone except Joseph Hannon.

I redoubled my efforts to woo him. Maurice of the green eyes and the gold-buttoned blazer had to fall in love with me. He had to like me because I would outshine all the girls behind the divider— do better in school, wear more striking outfits, be as fearless and accomplished as Mrs. Emma Peel. I thought that it was enough to be an agent extraordinary, to be the Avenger of Sixty-Sixth Street, to capture his heart.

I didn't understand yet that passion is fickle, that it is not possible to earn someone's affection, that love is not a meritocracy.

I was nine years old, on the verge of turning ten. Whenever I needed an infusion of courage, I repeated the Avengers' mantra out loud: *Extraordinary crimes against the people and the state have to be avenged by agents extraordinary....*

As little girls we had one enormous advantage. Adults tended to give us the benefit of the doubt, believing we were innocent and without guile. We weren't, of course—we were filled with dark thoughts and tormented by our own demons and obsessions that were every bit as terrifying as those that plagued our elders.

I often left my mother's side to huddle with my friends. She didn't really mind, but she always noticed when I was more restless than usual. "Loulou, reste ici près de moi," she'd say. Loulou, please sit still beside me.

But she couldn't dictate what I did.

Not here, not at the Shield of Young David. Though we enjoyed some privacy at the back of the women's section, I preferred retreating to the courtyard downstairs where we could play—and talk—in peace, without fear of being overheard by our mothers or by Mrs. Menachem, or chided by the men for making too much noise.

The concrete yard was the center of our universe. It was the common ground where four separate houses of worship, situated on two parallel streets, converged. The Congregation of Love and Friendship, where the newly arrived Egyptian immigrants, including Dad, prayed, was located on Sixty-Sixth Street, but had a back door that led to the courtyard. Its members, shy and poorly dressed, often strolled outside for some fresh air, as their little synagogue, housed in a two-family home, was overflowing with new arrivals. Next door was the Synagogue Without a Name, where the Eastern European Jews gathered in a grand majestic building. We'd heard that men and women regularly sat and prayed side by side—an arrangement that sounded both tantalizing and shocking.

But I never dared to go in to see for myself.

The courtyard also served the worshippers at the Big Shul next door to us, on Sixty-Seventh Street. That was the temple of the Syrian dowagers, an imposing building that dated back to the 1920s, with stained-glass windows and a domed ceiling painted with an azure sky. I occasionally wandered into the Big Shul simply to stare at its painted sky as well as at the rich older women in their mink stoles and gold bangle bracelets who sat in the balcony.

Most of the congregants at the Big Shul were from Aleppo, once the capital of Levantine Jewry. Syrian Jews had been dispersed all over the world, and wherever they went they made great fortunes and built large synagogues. The Big Shul was still thriving in the 1960s, though many of its members had moved away. Applying the lessons they'd learned in the Aleppo *souk* to the bargain stores and electronic shops they opened in New York, they made money and purchased imposing homes near Ocean Parkway ten minutes away. Those who stayed behind included many elderly who came faithfully week after week, the men in their impeccable suits, the women with their bracelets. It was said you could tell a Syrian woman anywhere in the world by the rows of gold bangles on her left arm.

The Shield of Young David was a newer, scrappier congregation, and none of the women wore mink. It drew Jews from all parts of the Arab world—Moroccans, like Madame Marie and her family; Yemenite Jews, including our cantor; some younger Syrian Jews, like Marlene and her siblings; a sprinkling of Turkish Jews; and a small Cairo contingent.

We were all Arab Jews, a culture most Americans found puz-

zling and that even other Jews viewed with suspicion. We had
no choice but to band together, and seek comfort and protection
among one another, shunning the world outside.

The courtyard was usually vacant, a kind of no-man's-land.
But I loved it and considered it as my own secret garden, albeit
one without any trees or shrubbery, only a vast expanse of gray
cement, where I could wander or have quiet conversations with
select friends. But during holidays, when services could stretch
from morning until evening without a break, worshippers from all
four synagogues would pour out for fresh air and linger chatting
in the courtyard and it became a social epicenter.

The disparate congregations never really mixed or mingled. It
was as if we were prisoners of where we chose to pray—the Syrian
aristocrats didn't deign to socialize with the new Egyptian immigrants, and no one ever spoke to the American Jews who attended
the Synagogue Without a Name. I never had a conversation or
played a game in the courtyard with an Ashkenazi child.

To us, the Americans were the ones who were foreign.

*R*abbi Ruben's daughters usually kept their distance from
the rest of us, but I made inroads with Miriam, the oldest,
a gentle, thoughtful girl. We would exchange favorite books, and
one day she lent me a thick tome she loved called *Rejoice, O Youth.*
It was an attack on the modern world written in the form of a
dialectic—an older rabbi answering the questions of his young
disciple—and it challenged among various popular notions Darwin's theory of evolution.

Rabbi Avigdor Miller, the author, argued that evolution was a
sham, a cruel myth. Darwin's theory was a "thick cloud of false-
hood," he said, a "fable." I was riveted by the back and forth between the rabbi and his student, which cast a contemptuous eye on

so much I was learning in school about prehistoric men and fossils and the like.

"What about the fossils, which the Evolutionist claims to be the remains of prehistoric men?" the student asks.

"There are none. . . . There are none whatsoever," the rabbi tells him.

It was all fine and good to believe men were descended from apes, but much of Darwin's theory rested on the "missing link," Rabbi Miller contended, the notion that there had been a creature who was literally both man and ape. But "Not a single one of these legendary 'Missing links' has ever been found," so where was the link, he kept asking.

I found Rabbi Miller's arguments stirring and persuasive, possibly more persuasive than what I was being taught about evolution in school. And so in the same way that at temple I questioned the need for a woman to get married, in school, I began challenging the sacrosanct theory of evolution.

I was being true to my contrarian nature, I suppose, when I raised my hand in class and demanded to know, "Where is the missing link?" I even argued with my teachers. In this battle for my soul that was waged behind the divider, this epic war to keep me and the other little girls safely shielded from the seductions of modern life, my assault on Darwin was unquestionably a breakthrough for the women of the Shield of Young David. While Mrs. Ruben worried about the toxic influence I could have on her daughters, the opposite was turning out to be true. I wasn't corrupting Miriam or her sisters—they were the ones who were prevailing, who were persuading me of the primacy of their ideas, at least when it came to Rabbi Avigdor Miller and his withering take on Darwin and the theory of evolution and the elusive missing link.

7

Passion Play on Sixty-Seventh Street

I was already known as a drama queen; now, I was going to be a star.

The women's section of the Shield of Young David was putting on a show, and I had landed a plum role.

At least, it seemed like a plum.

"Who is that dog of a dog who dares not bow down to me?" I thundered, standing in front of the mirror at home. I repeated the line again and again, emphasizing different words and trying hard to appear haughty and ruthless as I marched up and down the house. "Who is that dog of a dog who *dares* not bow down to me?"

We were reenacting the biblical story of Purim and I had been picked to play Haman *ha Rashah*—Haman the Evil One—the adviser to the king of Persia who decrees the murder of all the Jews of the kingdom, but finds himself thwarted and outwitted by the beautiful Queen Esther. In the last scene, I was to be marched off to the gallows, my head bowed in shame.

But that was only at the end. For much of the performance, I stomped around the stage hurling insults and commands and requiring everyone to kneel. I was sure that everyone in the congregation would take notice.

Of course, it wasn't everyone's attention I craved: There was only one person who really mattered. I was expecting Maurice to be there, of course. Marlene, his older sister, was directing the play and his younger sister Diana had a part. I figured I would finally have his undivided attention if not his heart.

I kept practicing and practicing my lines, while Suzette egged me on in my feverish need to sparkle and shine and stand out. Make the most of your lines she'd say during one of her frequent phone calls from Queens, where she lived in a small apartment with a roommate. Even if you have a minor role, even if you are only a little soldier, my sister would tell me, try to be the most important little soldier.

We had been rehearsing for weeks, and our play was the talk of the women's section since only little girls—the ragtag group of us from behind the divider—were being cast, tapped to play an assortment of biblical characters as well as courtiers and kings, ministers and magistrates. All of my friends were in it, and competitive as I was, I kept thinking that with my juicy role, I would steal the show.

I was dressed up to look like a Middle Eastern potentate. My long dark hair was hidden beneath a turban, or what I hoped would pass for a turban. It was actually a little brown knitted cap from Woolworth's on Eighteenth Avenue, but Dad had lent me swatches of gold and silver lamé from the big brown box of fabric samples he kept on top of the radiator, little bits and pieces of material he used for his new work as a textile broker. I had wrapped them around the hat, giving it a festive, regal look. My father was still a tie salesman, trolling the streets and subway stops for customers, but he was trying to build up a more respectable sideline dealing with the remnant stores of Lower Manhattan. Every day he walked from one fabrics shop to the next, offering to get them yards of silk or a shipment of velvet and showing off the samples that came in pretty little squares.

For my costume I wore my father's brocade dressing gown from Cairo. I had found it, neatly folded, in one of the twenty-six suitcases we'd taken when we'd left Egypt, several of which were around the house. It was in perfect condition; my father had never worn it here in America. But it was so immense that I had to hoist it up and tie the belt around me twice. Still, I loved marching up and down our apartment, Dad's old robe trailing to the floor, imperial in my yards and yards of sumptuous royal blue brocade.

As Haman, I was portraying a thoroughly sinister character, one of the most reviled figures in Jewish history. I reveled in the part when there, in the midst of my euphoria, I found myself plagued with nagging doubts: Why had I been chosen to play Haman? Why was I the one cast as a monster, a sadist of epic proportions?

Suddenly, the casting seemed rigged. My friends had landed far more sympathetic roles. Miriam, the rabbi's oldest daughter, had been tapped to play Esther, the lovely queen who fasts and prays and manages to persuade her husband, the king, to foil Haman's evil plot. Celia was playing Queen Vashti, the king's strong-willed first wife who defies her husband by refusing to dance naked in front of his ministers.

Mrs. Menachem, the show's producer, couldn't bring herself to say the word *naked*. She told us instead that Queen Vashti was summoned to appear in a "bathing suit." That only made us more curious, of course: What kind of bathing suit did a woman wear in biblical times—a bikini?

Gracie had landed the key role of King Ahasuerus, the ruler of all Persia. Her father, who owned a small factory on the Lower East Side, had sewn an elaborate costume of embroidered red velvet with gold trim that made her look stately and formidable. Marlene's sister, Diana, was playing a holy man.

Suddenly, I understood: This was Mrs. Menachem's little revenge. Marlene was directing the play in her usual earnest manner, but Mrs. Menachem was clearly in charge. And now, with this Jewish passion play, it was as if she had figured out a way to let the entire world know that I was arrogant and rotten to the core—exactly like my character.

In her eyes, I was increasingly lost to the outside world. I had fallen prey to forces beyond the Shield of Young David and its divider that were hurtful and corrosive and corrupt. I was becoming an American.

After the debacle a couple of years earlier, when my mother pulled me out of the House of Jacob girls' academy after a week, Mrs. Menachem had continued to push for me to attend a religious school. It was the only way, she warned Mom, to stop me from assimilating, to halt the process of acculturation that would take me away from all that she and my family and our small congregation held dear.

Mrs. Menachem was rigid and single-minded and extraordinarily focused. She was sure of her faith, sure of her traditions, sure of her ideas, and above all sure of what she wanted to avoid: any creeping influence of the outside world.

Hence her decision to put on a play coinciding with Purim. The holiday production was keeping me and all the other little girls busy and more connected than ever to the Shield of Young David. I was already at synagogue every single Saturday morning and most Saturday afternoons, and I attended the Hebrew school downstairs several evenings a week. Now, I was also coming for rehearsals. Any free time I had was centered around the compound on Sixty-Seventh Street.

And that, of course, was the point.

My mother found Mrs. Menachem and her worldview compelling. She'd come home from synagogue and muse out loud about the benefits of *la yeshiva*. She gingerly asked me if I would like to try the House of Jacob girls' academy again. Perhaps we had been too rash to give it only a few days.

But she never really pushed it. Her passion was reserved for the elegant private lycée and girls' schools of Manhattan, not a small parochial institution in Borough Park. If I couldn't be at the Lycée Français de Bab-el-Louk in Cairo, she dreamed of seeing me at the Lycée Français de New York. Deep down, the prospect of a religious Jewish school in Brooklyn left her cold.

Still, the fact that Suzette, her oldest, showed no desire to move back in with us filled Mom with despair and made her worry more about me. What was happening to our family in America? In Egypt, a young girl moving out was simply unthinkable. A daughter was expected to remain in her parents' home until the day she married, and this business of living on your own—so prevalent among American women—was horrifying to her, every bit as distressing as it was to my dad. That's why, more than two years after Suzette had left home, Mom still hadn't made her peace with it—with time, she seemed to grow even more upset.

Yet it wasn't that my sister had totally vanished from our lives. On the contrary, Suzette swooped down on Sixty-Sixth Street from her latest Queens high-rise every couple of months, often with gifts in hand. We'd feel so hopeful those afternoons as the conversation flowed and sparkled, and any sign of tension was gone.

Then she'd leave again and the house felt silent and empty—emptier than when she hadn't been there at all.

I never asked Suzette whether she'd be attending the play. I knew my sister could never overcome her revulsion of synagogues, and religion generally, even for the sake of watching me in my starring role.

Even on the major holidays, my sister didn't come. I'd set a chair for her at the table on Passover or the Jewish New Year, hopeful she'd put in an appearance for the lavish meal, but then when she didn't show up, Mom would lament that we were in a ruinous land and perhaps the only way to shield *pauvre Loulou* from ruination was to dispatch me immediately—immediately—to *la Yeshiva*.

On the other hand, Suzette loudly berated these religious institutions almost as evil cults, ready to snatch anyone who came within their grasp and brainwash them into dressing as they did, believing as they did, living as they did, marrying as they did.

She needn't have worried.

I had neither the soul nor the temperament nor the fashion sense of a *yeshiva* girl. I had no intention of wearing long skirts and long sleeves and a wig or getting married at eighteen. And the only cult I was susceptible to following was the cult of Emma Peel.

This was my time to be vain and flamboyant: My arrogant years had begun. I was extraordinarily conscious of what I wore and how I looked. I'd recently gone to our beauty salon around the corner on Twentieth Avenue with a picture of Mrs. Peel. This is the style I want, I told the startled beautician, handing her the photo of the Avenger, her lustrous auburn shoulder-length hair brushed to the side and away from her forehead. When I walked out, I was sure I resembled my heroine.

I wasn't a refugee girl whose parents didn't know how they

were going to pay the rent from one month to the next. I wasn't the witness to a disintegrating household. I was the fearless television seductress, taking on brutal enemies and defeating them with my physical and intellectual prowess even as I dazzled them with my wardrobe.

Suzette helped feed the fantasy. On my birthdays, she would arrive with a mountain of white boxes from Macy's filled with trendy, opulent clothes that conjured up Emma Peel and her effervescent 1960s fashions. We'd spread them out like treasures all over my bed—Carnaby hats, plaid miniskirts, jaunty vests, turtleneck sweaters—and I'd try them on one by one.

In the same way that I was meticulously planning my Haman costume, I was also fussing about what I would wear after the play, like some Broadway star preparing for her grand entrance at the after-theater party.

Mom cast a cold eye on my excess of vanity; she was capable of flying into a rage about my passion for clothes.

"Des tas de chiffons," she'd cry contemptuously when she was in one of her somber moods—a bunch of rags. She would urge me to please, please not be so frivolous. She was terrified that I was growing up superficial and self-absorbed. And she, the former beauty, the Ava Gardner look-alike who had captured my father's heart with her delicate brunette features, loved to tell me that beauty counted for nothing at the end.

I needed to study and read books and focus on my schoolwork because all the rest, particularly *ces chiffons*—wouldn't matter.

As if to set an example, she never wore makeup, shunned high heels, and had no use for jewelry. Her wardrobe was stark and simple and crushingly unassuming—a skirt and blouse, or a plain dress purchased from La Eighteen for a few dollars. I never saw her open the flask of Chanel No. 5 César gave her as a birthday gift, or wear the string of pearls he'd bought her when he'd felt extravagant and had a bit of money to spare from his earnings at Continental Grain.

For adornments, there was only the slender wedding band on her finger—no necklaces, no bracelets, no gold chains or rings or pins. What other women, even women of humble means, took for granted as part of the kingdom of being female, Edith shunned. There wasn't a single lipstick in the house, not one bottle of nail polish.

As if she weren't self-denying enough, the approaching Purim festival—meant to be joyous—was another fast day for Edith. She would be emulating the tender Queen Esther, doing without food and water to incur God's favor, pleading with him to void Haman's evil decree. My mother was always acting out her own sense of suffering and martyrdom on the stage of our family, starring as it were in her own passion play.

As the production neared, it became fodder for endless conversations with the other little girls behind the divider.

After prayers were over, I'd rush home for a fast lunch and then begin making the rounds of my friends and costars. First stop was Diana's house on Twenty-First Avenue, where I tended to walk each week, hoping to catch a glimpse of Maurice.

To my despair, he seemed unaware of me; those times he did acknowledge me, it was only in the most passing, distracted way.

Still, I would linger in my friend's cozy red-frame house, on the lookout for Maurice. My friend and I waited for him to appear as we chatted and played beneath an old antique coffee table, our little enclosure beyond the reach of the adults who came and went and generally ignored us, consumed with their important adult affairs. But now, with the play, it was as if the roles were reversed— our own lives had acquired an extra measure of zest, and the rest of the world, even this adult world we craved, seemed pale and lackluster.

Diana's home was a multigenerational household. Her grandparents lived downstairs while my friend's brainy, colorful family resided in a spacious apartment upstairs. Marlene was often around, and I'd chat with her about favorite novels we were both reading. I loved Linda, the oldest, the rebel, who never joined us at the Shield of Young David. She was edgy and tart-tongued and witty. The elderly grandparents were deeply moving figures who looked as if they'd stepped off the boat from Syria. One crucial member was missing—the father. Like my own sister's departure, his absence wasn't a topic of conversation. The rule among the women and children of the Shield of Young David was to avoid painful subjects.

Ours was a gossipy culture to be sure, but a rather gentle one. It wasn't like at the Big Shul, the old Syrian congregation whose members lived and died on rumors. There, a person's misfortunes became the topic of endless banter, peppered with requisite expressions of sympathy. "Hazeet"—poor fellow—they'd say, or its variant, "Hazeeta," poor girl, then resume trading sordid stories with even greater relish.

After visiting Diana, I would walk over to Celia's house. Madame Marie was always ready with a plate of freshly baked cookies and all the soda I could possibly drink. She outdid herself around the Purim holiday, and her platters of honeyed Moroccan delicacies—hand-delivered to select friends—were the talk of the congregation.

I was too nervous to eat; Celia and I spoke intently on the particulars of our roles and costumes. Her younger brother Moshe quietly watched us but was rarely included in our games or conversations. He wanted to be our friend, that was clear, but we ignored this sweet, awkward little boy who spoke with a stutter. He persisted in hovering around us, whether in his sister's room or on the edges of the women's section, and it was as if he preferred the intimacy of our little enclosures to the open space of the men's sanctuary.

My last and most important stop was to see the Cohen girls. Gracie and her sisters Esther and Rebecca were my safe haven, and I liked to end the Sabbath with them, journeying to their home on Twentieth Avenue. They lived in a small apartment on top of a storefront, all seven family members crammed into a handful of rooms. Their father, Abraham Cohen, owned the building and rented out the space downstairs to an Italian social club where mysterious groups of men gathered each day for card games and small cups of espresso and God knows what else, since 1960s Bensonhurst was known as a hub for the Mob.

Upstairs, order reigned; Mrs. Cohen ran a strict household. Each of the girls had to do what my own mother never asked of me: lots and lots of chores, from making beds to setting the table and washing the dishes.

The sisters were in constant motion, even when we were engaged in intense conversations about the play. In the midst of our chat, Gracie would be on her feet wiping the table or Esther would suddenly sprint to the kitchen to mash avocados for guacamole. There were two other sisters, Leah, the oldest, and Margarita—Maggie— the youngest, and both were "mentally retarded" in the parlance of the day. Leah seemed fairly engaged, but the little one, blond and delicate, was lost in her own world, a sweet helpless soul who'd wander over to us and smile, her head moving from side to side.

It was a close family despite the hardships. If Abraham and Adele Cohen suffered from having five daughters to raise, two of them disabled, they never complained. The father had a special bond with Maggie; the pair were often together, walking hand in hand. Because Mr. Cohen was a man of few words and Maggie was unable to say very much, they had a perfect entente.

I'd sit with the sisters in the main living area until the room became dark and the sun set and the Sabbath was over; only then did the Cohen sisters switch on the lights and life resumed once again.

After my sister left home, Mom became increasingly attached to Marlene. The lovely young girl was all that her own daughter wasn't: observant when Suzette had lost her religion, traditional when Suzette scorned tradition, content to sit beside us every Saturday behind the divider when Suzette would never set foot inside the Shield of Young David, let alone pray in its women's section.

"Marlene est un ange," Mom would exclaim. Marlene was an angel.

There was even an odd physical resemblance.

Both Suzette and Marlene had thick dark hair they wore in a pageboy, dark eyes, and a similar pale complexion. One afternoon, when I was on Bay Parkway, I stopped to examine the delectable chocolates on display in the window of Barton's Bonbonniere, the store that was very popular in our neighborhood because every single treat it carried was strictly kosher. I peeked inside the store and to my amazement I thought I spotted my sister standing by the cashier.

Was she paying us a surprise visit? Was she bringing us a gift of chocolates from Barton's? I waved frantically and started crying, "Suzette." Then I realized to my confusion that it wasn't my sister at all; it was Marlene doing a bit of shopping.

Marlene had been coming to the Shield of Young David since she was a little girl, when the synagogue was new. It functioned as an annex to the crowded Big Shul next door. Of course, the elder Syrians, those aristocratic men from Aleppo and Damascus whose wives came dripping in furs, gold bracelets up and down their arms, had been praying at the Big Shul for decades. They weren't going to budge, but they welcomed the chance to send their children and any overflow crowd next door.

Marlene felt at home there from the start, and the synagogue became her playground. She and the other children would retreat upstairs to the roof during services where they'd run and play and jump high above the rest of the neighborhood. No one sought to warn them of the dangers and, indeed, none of them felt any danger. That was the wonder of the Shield of Young David: It always seemed safe, even on the rooftop.

A pensive little girl who wasn't exactly a daredevil, Marlene found herself joining her friends in a dangerous, potentially lethal game they played several stories above the ground. It was a far cry from her role of dutiful older sister, expected to shepherd her younger siblings to services each week. Now, though only a teenager, Marlene was like our elder statesman. She was among the few to enjoy the friendship of the rabbi's wife, the trust of Sarah Menachem, and the affection and respect of rank-and-file members such as my mother. Despite her quietness, she was also opinionated, never shy about letting us know her views.

On the eve of the Zero Population Growth movement, when the notion of having small families began acquiring cachet, I promptly told everyone in the women's section that we should all have at most one or two children. Marlene coldly, immediately disabused me of the idea. Jews had lost so many of their own in the Holocaust, why not "let others" pursue the noble goal of population control?

I couldn't even think of a comeback.

As our performance drew near, Marlene emerged as a stern, formidable taskmaster, barking orders, requiring us to practice our parts again and again, and then critiquing our individual performance. Had we memorized our lines? Were we going over them at home? Had we organized our costumes?

I felt feverish those last few days. Even in school I found myself repeating my part over and over again, my mind drifting from whatever the teacher was saying: "Who is that dog of a dog WHO DARES NOT BOW DOWN TO ME?" At the Shield of Young David, any stray men who happened to be in the sanctuary praying or studying during our rehearsals knew to get out of our way.

It felt like a historic event—our synagogue putting on a play. Sunday afternoon, women and men filed into the main sanctuary together and took seats side by side. I had never seen that before.

And then we began.

I made my grand entrance, trying hard not to trip on my father's long brocade robe while I balanced the silk turban on my head. I had been made up to look fierce and angry—one of the women had drawn a mustache in thick black ink and made my eyebrows dark and bushy.

"Who is that dog of a dog who dares not bow down to me?" I roared as I marched onto the stage. The audience burst out laughing and applauded and booed at the same time. "Who is that dog of a dog who dares not bow?" I repeated and pointed to Mordecai, the lone Jew who refused to yield to my command.

We had no curtains, no sets, no furniture onstage, yet our audience seemed enthralled. They all shared the illusion that we were back in biblical times. One by one we walked onto the stage. Gracie, splendid in the velvet outfit her father had stitched painstakingly by hand, looked every inch a king. I cowered when I was denounced as a madman and traitor, and tried in vain to protest when Gracie revoked my decree to execute all the Jews in the kingdom and ordered my hanging instead.

I was led away from the stage, my head bowed in shame.

Once the play was over, I tore off my robe and turban and rushed back to the sanctuary. Everyone seemed ecstatic, congratulating each other, hugging each other, and even Mrs. Menachem was beaming, the edge temporarily gone from her. "Loulou, tu était absolument magnifique," Mom said—I had been magnificent.

I glimpsed Maurice in a corner and kept trying to catch his eye. I was a star. Surely, this of all days would be the day I would have earned his love, the day when he finally noticed and acknowledged me.

He didn't, of course. He simply left and went home along with everyone else, and I wandered around the sanctuary searching for the sense of achievement that never really came, no matter how hard I worked for it, no matter how hard I tried to obtain it, the attention that I craved and that continued to elude me again and again.

8

The Healing Powers
of Jodine

ost everyone at the Shield of Young David vanished the moment school ended and summer began. My friend Diana left for the Jersey Shore with her mother and siblings, including Maurice. Other children fanned out to the Catskills or else to sleepaway camp. The aristocrats in our congregation traveled to Israel.

We were left to vacation in Bensonhurst.

But to do what?

Looking out onto the men's world beyond the divider Saturday mornings wasn't nearly as riveting with Maurice gone. At home, my mother seemed daunted by the prospect of keeping an energetic, high-maintenance child busy and entertained over an entire grimy New York summer.

It wasn't like Cairo where the family packed up each year for Alexandria or what Mom dubbed our "grandes vacances." There was no family left anymore, not really. My siblings increasingly were going off on their own. Suzette was ensconced in one of her ever-changing Queens high-rises. I'd receive letters from her with return addresses that sounded both distant and exotic: Lefrak City, Rego Park, Forest Hills, Kew Gardens. My brothers were still with us, but they had their friends, their own lives really, and were rarely at home.

This was all terribly painful for my mother, who felt she was watching the family disintegrate yet seemed helpless to stop it. What she did instead was to hold on to me even more tightly, to focus on me more intently.

She was still reeling from a year of tangling with my public school, the fact that teachers and administrators had balked at her demand to have me skip a grade, since she was convinced I was wasting my time at PS 205.

Edith was also upset that there was no prospect of *grandes vacances*. In the summer of 1966, only she and I and dad were left, and we certainly couldn't afford to go to Alexandria's American counterpart.

Not that we thought there was a counterpart, not that we believed Miami Beach or Bradley Beach (where the Syrian Jews were congregating) or any resort in this country could possibly compete with the yellow sand of Agami Beach, so fine it was like talcum powder to the touch, or the water at Mahmoura that was limpid and clear and you could look down and see the pebbles at your feet, or the lively Corniche along Stanley and Sporting beaches with its dozens of charming cafés serving pitchers of beer and *mezze*, deliciously cold appetizers that you could munch on while facing the sea.

Mom's strategy was to have us go to the poor man's Alexandria—Brighton Beach in Brooklyn.

The morning after school was done we boarded the Sea Beach

Express, then changed for another train in Coney Island. Once we reached our stop, we clambered down the steps where we found dozens of fruit stands and kosher butchers and small bargain stores crowded together in the dark careworn spaces under the elevated tracks.

As we walked to the beach, we passed a private gated club known as the Brighton Beach Baths. I'd peer through the metal fence and see elegantly dressed men in white shorts playing tennis and women in bathing suits lolling about by large swimming pools or soaking up the sun on comfortable folding chairs. I had eyes only for the miniature golf course on the other side of the fence. It was so alluring, with its little windmills and minicastles—a make-believe kingdom sized for a child.

The prior summer we had stopped at the entrance to inquire about admission. Whether we asked about the summer rate or the monthly rate or the weekly rate or even the price of a one-day pass, the Baths were completely beyond our means and I had to content myself gleaning what I could from the holes in the fence, much as I followed the goings-on in the men's section at the Shield of Young David through the apertures of the wooden divider.

On the other side of the road leading to the beach, even closer to the ocean, was a summer camp operated by a Y. I glimpsed groups of children my age laughing and frolicking in a large swimming pool. They all wore little yellow plastic caps and I found these caps as well as the notion of camp, of an organized system of games complete with playmates, both foreign and appealing.

My longing to attend the Y day-camp left Edith cold. Summer camp in Brooklyn? The notion of letting children swim in a chlorinated pool when the ocean was steps away struck my mother as absurd, a quintessentially American folly.

Every once in a while, my mom would wistfully describe how she wished I were really spending my vacation. She'd invoke our

Milanese cousin Salomone who sent his children every year to sleepaway camp in the Swiss Alps.

A camp in Switzerland on the shores of Lake Geneva: Now that was a worthwhile destination for *pauvre Loulou*.

Edith's options for me were always so fanciful and out of reach. It was a miracle that she managed to scrape together the subway fare to get us both to the beach every day, let alone send me off to a fancy Swiss camp.

Her musings about an Alpine vacation for me reminded me of her equally hopeless quest to have me skip a grade. Her battles with my teacher, my principal and, finally, the bureaucrats at the Board of Education had dominated the better part of the last year. At one point, my fourth-grade teacher, Mrs. Simon, an attractive blonde with a dry wit, wrote to say that while I was a lively, engaged pupil, I "should not be pushed." My mother was stung. Instead of graciously accepting the criticism, she fired back with a note of her own. "She is not being pushed. She *wants* to learn."

The fact was that, for Edith, the idea of her daughter attending public school was simply anathema. Indeed, from the moment I had entered PS 205, she had made it clear she despised the place— the quality of the education was nothing like what I'd experienced at the Lycée Français in Cairo. My studies weren't nearly intense enough nor were my teachers rigorous enough to satisfy her boundless ambition on my behalf.

During the year, to supplement what I was learning—or more likely not learning—she would give me private lessons. Every afternoon, we would sit down together and she'd try to teach me French grammar, or give me *des dictées*. Those were elaborate spelling tests, central to the Parisian educational model Mom revered. She'd read out loud a long and often difficult passage from literature—not simply individual words—and I had to write it down.

Afterward, she'd carefully scour my *dictee* for spelling or grammatical errors.

The star teacher of L'École Cattaui, the disciple of the pasha's wife, went over key episodes in French history, introduced me to the classics of French theater, and then, appalled at the rudimentary arithmetic I was learning at school, added math to our curriculum as she began teaching me advanced multiplication tables, long division, and percentages.

Then there were the pop quizzes—the questions that she'd ask me at any point in the day, even as we strolled to the beach. What century did Louis XIV reign? What was his nickname? When were the plays of Racine and Corneille performed? What year did Flaubert publish *Madame Bovary*?

If she was ever dismayed by my performance, she didn't let on, even as I got answer after answer wrong, mixed up the masters of the seventeenth, eighteenth, and nineteenth centuries, and couldn't for the life of me distinguish between the great tragedies, so that I confused Corneille's *Medea* with Racine's *Phaedra*. I hadn't the foggiest notion when Flaubert had published *Madame Bovary*, the book that Edith told me again and again, even when I could barely read, was the greatest work in the history of literature.

Edith was intensely patient with me, but she could never stop raging against PS 205. She was on a warpath—in meetings with the principal, she'd informed him that American schools were decidedly inferior to French lycées; I had to skip a grade, ideally two. I had to be with more advanced pupils.

Her appeals—that I was a child who needed to be stimulated—landed on deaf ears. In the rigid bureaucratic culture of New York City's public schools, the liberating influence of the 1960s was being felt, though not necessarily for the better. The prevailing liberal ethos was that all children were equal and had to be with their peers. Age, not talent or knowledge or ability, was what mattered most.

I was to be educated with all the other nine-year-olds, we were told repeatedly.

Defeated by the system, humiliated by its army of bureaucrats, Edith was thrashing out a new strategy for my education as the summer of 1966 began: I was going to attend private school exactly as I had in Cairo.

But not simply any private school.

Edith fixed her sights on the Lycée Français de New York. A supremely tony institution on Manhattan's East Side, it catered to an elite clientele—children with addresses that didn't stray far beyond the contours of Fifth and Park and Madison. Tuition alone was steep, and it was fairly obvious that connections were needed to get in, not to mention wealth, social class, and a rigorous knowledge of French.

Yet Mom latched on to the dream of seeing me once again as *une lycéene.* To her mind, her children had always attended the lycée and while our circumstances had changed dramatically, she still felt that I belonged there.

She imagined me going off to school in the magnificent mansion on East Seventy-Second Street off Fifth Avenue, carrying my books in a neat leather satchel—not clasped together with a large rubber band as my uncouth public school classmates did—and wearing a uniform.

Among Mom's many pet peeves, she loathed the fact that public schools allowed children to wear whatever they wanted to class. It was, of course, another one of those egalitarian American impulses—uniforms like private schools were so elitist, we'd be told again and again. But Edith had her doubts. She wasn't sure America was that egalitarian. Uniforms were the true equalizer—rich or poor, all children were forced to dress the same. Besides, how else to outfit a child for school, without having to fuss every day or spend a lot of money for a wardrobe? The Lycée Français had a uniform, of course, an exquisite one, a navy blue blazer over a pleated skirt that she considered the ultimate in elegance and style.

Clearly, the only hope for me was to rejoin the company of *des jeunes filles de société,* the upper-class girls she felt were my peers and not those working-class kids, the children of seamstresses and sanitation workers and cabbies, who were my schoolmates now.

One of the few luxuries Mom allowed herself was a subscription to the *France-Amerique,* a French weekly newspaper. Only a few pages thick, its readership consisted of an upper-crust base of French expatriates, members of the diplomatic corps, and Francophiles. Any French cultural event in New York, any concert or play with a remotely French theme, any French artist was sure to be featured.

But what Mom treasured most of all were the ads for the lycée and the occasional news items about it. She scoured the paper for insights into the school, noting the powerful people connected with it, and formulated her plan of attack.

She would approach the lycée directly on my behalf, she decided. She would mount a full-fledged campaign by sending its president as well as a raft of other officials her exquisitely crafted letters, imploring them to admit me and to offer me a scholarship.

Her strategizing would start now, on these early July outings to Brighton Beach. She had decided to devote the summer to figuring out how to get a place for me in that most elite, desirable of schools.

But first we had to get our supply of iodine and build up our defenses for the battle ahead. Besides, the ocean beckoned—not like the sea at Sporting, perhaps, our favorite beach in Alexan-

*Edith with twelve-year-old Loulou at
the beach, summer 1968.*

dria, but still so appealing, and as Mom reminded me, chock-full
of iodine.

My mother had a mystical belief in the healing elements of the
sea, its powers to cure any malady. The operative ingredient was
iodine—*l'iode*—the compound she believed was so potent that it
transcended most medicines. Every summer, she'd advise me to
immerse myself in the water and remain there as much as possible.
Our trips to Brighton Beach weren't merely supposed to be entertaining but therapeutic, a chance to wash away any incipient bug
we could be harboring, alleviate a chronic condition, even treat a
serious malignancy.

In despair over her unsightly varicose veins, Edith was convinced that after a good dip in the water, the bulging lines at the
top and the back of her legs would magically recede or disappear.

A believer in alternative medicine years before it became fashionable, she would inspect herself closely before each of our swimming expeditions. Then, after floating in the ocean for a while, she would turn to me and ask me anxiously if her legs looked any better.

I would never think of questioning her faith in *l'iode*, nor did I have the heart to disappoint her. I'd peer closely at the spot where she was pointing and nod that yes, the veins were fading away, no question, and she was overjoyed.

In Egypt, when we had access to the Mediterranean only for a certain number of weeks, she and Leon were both so worried about getting their necessary doses of seawater that they would actually pour some in bottles at the end of August. The bottles would be sealed, packed away in our suitcases, and carted from Alexandria back to Cairo to help them weather the year. When his eyes bothered him, for example, my father would take out one of the bottles, pour some seawater on a cotton pad, and apply the soaked pad to his eyelids. Mom liked to wash her face with the seawater.

In Brooklyn, my parents' worries about access to iodine abated—the beach was only a subway ride away, after all. And while each of them found this country was missing so much of what they had cherished about their lost life, they agreed that the Atlantic was every bit as rich and wonderful and dependable a source of iodine as their beloved Mediterranean.

When Dad joined us for a day at the beach, he would wear shorts or roll up his trousers to reveal his ulcerated left leg. It had been nearly a decade since he'd suffered that terrible fall that had shattered his leg and hip. The accident had taken place in Cairo shortly after I was born, and he had never quite recovered; on the contrary, he was always in discomfort and no prescription ointment or cream seemed to soothe his injured leg.

But then the iodine came to the rescue. Both were hopeful that the water—far more than the doctors' salves—would at last heal Dad's painful leg. Mom would take my little sand pail and go fill it with seawater that she'd pour over his affected leg in a move that struck me as supremely tender.

Because I was prone to so many maladies—the victim of summer colds, mysterious bugs, hacking coughs, fatigue, fevers— my mother would plead with me to scrub my face and my hair in the Atlantic and let the sea wash over my entire body. She usually went in with me, joyous in her old-fashioned one-piece blue stretch bathing suit that she had brought from Egypt and still insisted on wearing. She never spoke about the darker side of iodine or discussed that horrible incident at Villa Lalouche in Alexandria before I was born.

And while I never thought of my mother as especially carefree, those moments in the water when she and I together made clumsy attempts at swimming, jumping up and down in the waves, I had a sense she had finally let go of the angst that seemed to perpetually grip her, that her demons had at last relinquished their hold. There in the water, it was as if she had never really grown up—she seemed in many ways a little girl, a waif, a giddy, excited child, not too different from me.

Afterward, we settled back on the little blue kitchen towels we had brought from home. They were so small, we couldn't really lie down on them; but that was fine, because we were both too restless—neither Mom nor I were the type to stretch out and bake in the sun. We preferred to sit up and enjoy the amiable scene around us—families devouring large meals of chicken and sandwiches, young girls parading in their skimpy bikinis, lifeguards preening up and down, even as portable transistor radios blasted Cousin Brucie, the relentlessly cheerful DJ who epitomized a Brooklyn summer, and soda vendors with large coolers strapped to their shoulders trudged through the sand pleading to sell us a cold drink.

*W*e left the beach the way we came, retracing our steps past the children of the Y and the dapper members of the Baths.

In 1966, Brighton Beach seemed dominated by elderly Jewish women, including many Holocaust survivors. These had moved to the area decades earlier and raised their families a few steps from the El, in humble walk-ups and tenements that had breathtaking ocean views. But now their upwardly mobile children were gone, dispersed to fancier sections of the city or the country, their husbands had passed away, and it had become a neighborhood of widows, or so it seemed as we walked. We'd pass groups of aged women sitting on folded beach chairs, chatting outside their dilapidated buildings. They'd be sunning themselves, uninhibited about sitting on the sidewalk in bathing suits that revealed all their fat and wrinkles.

How we envied them. "Wouldn't it be wonderful to live close to the sea?" Edith would always exclaim. And that was her most fervent dream—that we would move to Brighton Beach and have access to healthful iodine.

As we turned the corner to Brighton Beach Avenue and its bustling shopping strip, I felt the fun was only starting.

I shared Edith's fascination with the discount stores beneath the El, the shops that displayed their wares in vast, inelegant bins. If Mom missed the fine elegant shops of Cairo, she rarely let on; only once in a while did she speak wistfully of Cicurel, the large department store on King Fouad Street that sold only the most exclusive merchandise in Egypt.

In that sense, she was different from Dad, who couldn't abide bargain stores and cheap merchandise. Unable to afford the fine clothes he had always loved, my father chose not to shop at all. Only once in a while, in small extravagant gestures reminiscent of his old self, did he come home with an elegant straw Panama from an upscale Manhattan hat store.

Mom and I felt differently. In our eyes all the promise of America was there in emporiums such as John's Bargain Store, where for small sums of money, it was possible to decorate an apartment, assemble a wardrobe, purchase gifts. Best of all, we could lose ourselves in those bins. There was such a hopeful feeling that came with rummaging through the boxes and containers filled with items on sale.

Mom left me to go explore the houseware department—containers overflowing with window curtains, towels, throw rugs, sponges, bedding, saucepans, and gleaming faux-silver flatware that cost as little as ten cents for an individual fork, spoon, or knife. Shopping for a small carpet, or even a spoon, made Edith hopeful.

She was always quoting a line from Alphonse Daudet, one of her favorite novelists, "Il faut reconstruire le foyer."

It meant, "You must rebuild the hearth."

Mom was obsessed with the notion of reconstructing what we had left behind on Malaka Nazli Street. She longed to pull our family back together again, to re-create a semblance of the home that we had lost, and it was as if she believed she could do it bit by bit, piece by piece, with pillowcases, towels, and spoons.

Next stop: the shoe store across the street. Summer began with another tradition that dated back to Egypt—the purchase of white sandals. In settling here, we'd had to eliminate our habit of wearing white—most especially white shoes. Westerners didn't wear white shoes, and putting them on would have branded us immediately as foreigners or, worse still, immigrants. White shoes in America were for proms or weddings or confirmations, not everyday life. The many pairs we'd brought with us lay fallow in the bottom of our twenty-six suitcases. We were almost afraid to open them, afraid of what we would find from our lost life.

The one exception was summer, where it was appropriate at least for children to wear white.

Off we'd go to Miles, the cool, darkened shoe store near the beach. Inside, quiet and order reigned, and the children's sandals were expensive; there were no markdowns, no sales, no bins—these would come only at the end of the summer, the salesman told us, when prices would be slashed by half or more. But by then, of course, we wouldn't want them.

Mom had decreed I could have one new pair of sandals, *des sandalettes*, as she called them. They had to last me through July and August and daily walks on the boardwalk and in the sand and they had to be white and they had to be made of leather. I'd try on as many pairs as the salesman would allow and choose carefully.

Changing into the new sandals seemed to have a transformative effect. As the obliging salesman strapped them on and I tossed away my dark wintry school shoes, I felt as if an enormous weight had been lifted. I could run, skip, jump, almost float in my white sandals. I could revert to an earlier incarnation of my childhood, when I was a little girl in Egypt and I didn't shop in bargain stores and white was all I ever wore.

Light and carefree in my *sandalettes*, I headed with Edith to the kosher butcher shop with the intriguing neon Self-Service sign on the window that heralded a different kind of establishment than any we'd ever known. Modern, antiseptic—it wasn't at all like the old-style butcher shops we patronized in Bensonhurst (or for that matter in Cairo) where men in bloody aprons chopped and prepared whatever meat we ordered on the spot. The shop consisted entirely of freezer sections where you could find impeccably wrapped packages of ground beef, steaks, lamb chops, and hamburger patties at the exact weight and price you wanted. The butchers themselves were nowhere to be seen.

Back home we collapsed from exhaustion over the iodine and the shopping expeditions. We'd sit together and have a simple meal Mom prepared, typically scrambled eggs with tomatoes and green peppers.

It was over one such after-beach dinner that my mother unveiled her grand plan for my future.

I wasn't going to go back to "cette affreuse école public"—that awful public school—she declared. She was prepared to yank me out of PS 205, and beg and implore the lycée's formidable president, Monsieur Galy, to admit me. If that failed, she had another ace up her sleeve. She would reach out to Edouard Morot-Sir, the influential cultural attaché at the French Embassy in New York.

Surely, if anyone could get me into the lycée, it would be the diplomat with the aristocratic hyphenated last name.

She wasn't being entirely fanciful—my family had, against all odds, forged a relationship with Morot-Sir. Edith had begun corresponding with him shortly after we'd arrived in America, when my brother and sister needed to have their studies in Egypt accredited to be able to attend college here. To her surprise, the eminent diplomat—whose name she always pronounced in a reverential way—responded amicably. Indeed, Suzette received two years of college credit as a result of his help.

The illustrious Edouard Morot-Sir would be a formidable weapon in Mom's arsenal. He would be her way, and mine, into the elusive lycée. She reiterated her disappointment with PS 205. Even now that I was trying to put the year behind me, she was still brooding and obsessing over it and figuring out how to implement her game plan.

ℰ dith became so absorbed with her letter-writing campaign to the Lycée Français, that the daily outings to the beach decreased and she left me on my own, expecting me to find ways to keep myself busy.

I must have seemed disconsolate. One Saturday my mother turned to Madame Marie, her closest friend in the women's section. She was like us—she couldn't afford to go away. The two agreed both Celia and I were in need of a respite. There we were, rows of empty chairs all around us, and no other children in sight.

Their solution?

We would go on a cruise—a cruise to Staten Island.

When Madame Marie suggested a boat trip to Staten Island, Mom readily agreed, though neither of us had the foggiest notion where or what Staten Island was. Since we had arrived in America I had rarely wandered beyond the confines of Sixty-Sixth Street, though I was hungry for adventure.

Sunday morning, we found Madame Marie, her husband, Celia, and younger brother Moshe ready and waiting for us in their cramped apartment on Sixty-Ninth Street, and together we set out on our journey. A magical cook whose dishes were redolent with the spices of her native Morocco, Celia's mom had prepared a big picnic for us.

Once at the ancient South Ferry station, we clambered aboard the boat, giddy at the prospect of a great sea voyage. We had no idea how far Staten Island was—it could have been hundreds of miles away for all we knew. Perhaps we were going on one of those *grandes vacances* after all.

Once the boat docked, the question arose—where to now? I saw Mom glance at her friend with alarm.

We were strangers in a strange land. We may as well have been standing on the dock of a foreign port. Edith took charge, bravely asking a passerby if there were a nice park nearby, "to take the children." The man directed us to a bus and we hopped in feeling terribly adventurous.

There was a dreamlike aspect to the ride; we looked out the window at a landscape that was startlingly different from the rows of two-family homes and brick tenement buildings we knew in Brooklyn. Here, we passed vast open spaces, the occasional neat private house with a lawn, and water all around.

We finally arrived at a lush green wooded expanse. "Clove Lake," the driver barked amiably. We walked around in a daze to a park that bore no resemblance to any we knew in the city. It looked and felt like an enchanted forest, with thick trees giving way to an occasional clearing.

We settled onto a grassy patch and prepared to dine. Madame Marie had outdone herself. Instead of a picnic basket, which neither of our families owned, she had brought several shopping bags filled with her specialties. None of us wanted to leave—I felt as if I could have stayed forever on this strange idyllic island. As we boarded the bus back to the ferry, we agreed that Staten Island was full of mystery and we had to find a way to come back.

9

The Errant Sister

Although Mom and Suzette clashed on every possible subject, they were usually in agreement about me. When it came to dissecting the American school system, they were kindred souls—public schools were atrocious, my sister remarked when Mom briefed her on her plan to take me out of PS 205. My mother found in Suzette the perfect ally—someone who shared her sweeping ideals as well as her grand illusions. Suzette firmly concurred that I should be given a shot at a decent private education—the one so cruelly interrupted when we left Cairo and I had to abandon the Lycée Français de Bab-el-Louk. Only the Lycée Français in New York would do.

Yet even with a subject they could discuss harmoniously, Mom couldn't resist needling Suzette. Her rage at my sister and the fact she insisted on living away from home manifested itself in nearly all their dealings. While Edith was more progressive than my father and had always cut my sister some slack, it was now two years since Suzette had moved out. There was no sign she was coming back, yet Mom could never stop pleading with her to return. And though my sister only lived in Queens and stayed in constant touch through phone calls and letters, her departure haunted the family, overshadowing each of our lives. It was the great calamity of our move to America: We had lost Suzette.

Every phone call to my sister, however peaceable, seemed to end with Edith delivering a tirade.

This time it was about how Suzette was neglecting me, "cette pauvre petite enfant"—this poor little child. Mom told her I was struggling in this difficult New York summer, and she was finding it hard to keep me busy, and what was the point of having an older sister anyway? Suzette slammed the phone, and my mother began to cry. What had been a cordial conversation about the horrors of the American educational system in general and the advantages of the lycée had turned into another bruising battle.

Mom's words must have penetrated. Within days of this latest clash, my sister materialized, ready to take me to the beach.

Even back in Egypt, Suzette never liked going to the beach with us. Sporting Beach, located in a middle-class section of Alexandria, was a favorite destination of many of our friends and neighbors from the Jewish community. My sister couldn't stand it. Instead, she wangled invitations from her tony friends to Sidi Bishr or San Stefano, the far more exclusive enclaves farther up the coast, where the crème de la crème of Egyptian and European society gathered.

To Suzette, we had once been hopelessly middle class. Now, we were miserably working class. It had been relatively easy in Egypt for a young woman with charm and appeal (and my sister had plenty of both) to climb up the ranks and be embraced everywhere, from the elite private clubs of Cairo in the winter to the elegant seaside playgrounds of Alexandria in the summer. But here in America, ostensibly the land of upward mobility, life seemed

Eighteen-year-old Suzette on her French travel documents, Paris.

more rigid and castelike. Stuck in a low-level clerical position at the First National City Bank in Manhattan, my sister's future, instead of opening up to limitless possibilities, seemed more limited than ever.

She was twenty years old.

Yet her friends from Cairo, after a similar turbulent exile and relocation, were beginning to settle down. Her closest companions from the lycée, the Wahba sisters, had moved from Cairo to Paris, and finally to Israel. Pretty and charismatic, they were getting married and one by one, having children. They felt an optimism about the future that my sister, and for that matter, the rest of my family, lacked.

My mother was always citing the Wahba girls as shining role models. When was Suzette going to come to her senses? When was she going to resume her education? When was she going to get married? When was she going to come home?

My sister would always get so distressed, and the arguments would begin again.

hen Suzette showed up Sunday, she was determined to offer me, as Edith had urged, *quelques distractions*, a little bit of fun. Yet her snobbish tastes and condescending attitudes were on full display. We weren't taking the Sea Beach Express, she announced. We were going to Manhattan Beach.

Manhattan Beach, contrary to its name, was also in Brooklyn. But it was far away, and getting there by public transportation was complicated. We boarded a tangle of buses that took more than an hour to get to the only beach in New York that met Suzette's standards.

Once we reached the sand and prepared to settle in, I took out the faded little blue towel Mom had packed for me. My sister looked at it with contempt. She immediately marched me to one of the vendors hawking their wares by the entrance and bought me my first beach towel—massive, soft, big enough to fit two people side by side. We stopped at another stand where I picked out some toys—a blue-and-white inflatable ring and a beach ball. She also grabbed a bottle of Tropicana suntan oil—another luxury my frugal mom would never have considered buying.

At last, we were ready to proceed to the ocean.

Manhattan Beach was exceptionally crowded. Its topography seemed different to me from Brighton Beach, and I felt disoriented. But Suzette was determined to make me feel at home there—offering me suntan lotion and applying it on herself, like the models on the TV commercials. My sister looked especially glamorous in her yellow bathing suit, which set off her dark hair, and she stood out even among the bikini-clad teenage girls. She had brought a novel to read and encouraged me to take my little inflated ring and enjoy myself "dans cette belle plage," on this lovely beach. Then, she settled into her book and her suntan.

I ventured out rather nervously. The water seemed to get deeper much faster than I expected.

I tried to practice my swimming using the ring. I was so absorbed floating on my back in my little ring that I didn't notice the group of boys who had formed a circle around me. They were laughing, and one grabbed the blue-and-white ring and began pulling me deeper and deeper out to sea. I tried to scream, but because of the waves and the sun and the crowds of people around me nobody could hear me. The more I cried, the more the boys laughed.

Where was my sister?

When they finally scattered and left, I was far out into the water. There was no one around me, and I was sure I was going to drown.

At last I spotted Suzette swimming toward me. I was crying hysterically when she reached me. She seemed mostly bemused, asking what had happened. She pulled me back to shore and together we walked back to our encampment and my new beach towel. For the rest of the day, I hovered close to her on the sand.

I didn't tell Edith about what happened. Back home when Mom asked "Comment était la nouvelle plage?"—How was the new beach?—I said only that it didn't seem nearly as nice as ours. The sea didn't even have that much iodine out in Manhattan Beach.

My mother looked at me strangely, but she didn't press me. I couldn't tell her the truth. She would simply have added it to the ever-growing list of sins my sister had committed against God and the family and me. She would have condemned Suzette as selfish and hopelessly self-absorbed. She would have reminded her that since she'd left home, my brother was supporting the five of us on his slender salary from Continental Grain. She would have demanded to know how she could have possibly let *pauvre Loulou* out of her sight at a strange and faraway beach.

10

The Passion of the Fast

S uddenly, summer didn't feel much like a vacation. We had entered the period of lamentations.

I wasn't allowed to buy new clothes or listen to music and, worst of all, I couldn't go swimming at Brighton Beach. Toward the end of July, life changed—it wasn't about taking in iodine anymore. It was about recalling ancient tragedies dating back thousands of years. It was about reliving the capture of Jerusalem. It was about fasting.

It was about self-sacrifice and suffering—lots and lots of suffering.

Edith was an old master at that.

Indeed, whenever I hear the story of Rabbi Tsadok, the holy man of ancient Judea, I am reminded of my mother in our old kitchen on Sixth-Sixth Street, gearing up to observe the festival of lamentations. Rabbi Tsadok became famous for fasting forty years to avert the destruction of the Second Temple. He failed, of course. The temple was reduced to ruins, Jerusalem was captured by the Romans, and Rabbi Tsadok himself became painfully thin and frail. Yet to scholars and historians, he remains a hero because for those forty years, the temple stood and Jerusalem was safe.

Rabbi Tsadok is perhaps the only person who matched Mom's passion for fasting. Edith approached the act of going without food and water with an abandon I never saw in anyone else. In the course of my childhood, I noticed that no fast on the Jewish calendar was too minor or trivial or obscure. While most of our American Jewish friends fasted only once a year, Edith at times fasted almost once a month.

There was, of course, the Fast of Gedalia, when she mourned the assassination of a governor of Judea in 582 B.C., or the Fast of the Firstborn, when she recalled the final plague against the ancient Egyptians, that night the Angel of Death traveled from house to house to slay all their firstborn sons. The fast was typically mandatory for firstborn sons, but among Syrian and Egyptian Jews, women were also expected to fast and Edith, a firstborn daughter, did so rigorously. She fasted on the eve of Purim, a holiday of parties and gifts and costumes that begins sadly, with a day of soul-searching and no food or drink. She observed the fast of the Tenth Day of the Tenth Month, when the Babylonian king encircled Jerusalem, as well as the Fast of the Seventeenth Day of the Fourth Month. That fast was to mark a day of unremitting grief—the day Moses came down from the mountain and caught the Israelites worshipping the Golden Calf, the day he smashed the tablets containing the Ten Commandments, the day when all sacrifices to God ceased because there were no more animals left to slaughter, the day the sacred Torah scrolls were set on fire, the day an evil king placed an idol inside the Great Temple and desecrated it, the day the wall around Jerusalem was breached and the enemy army entered, and all hope was lost.

Mom's fervor manifested itself most dramatically at the height of summer. Even as the rest of the world was reveling in all the pleasures the season brings, we had to get ready for the Fast of Lamentation, when we mourned the loss of the temple that Rabbi Tsadok had tried so hard to save.

It was such a somber event, it literally required three weeks of preparation. I watched Edith change before my eyes.

It wasn't that Mom was incapable of great *joie*—on the contrary, I was always struck by her girlish bursts of enthusiasms, the fact that she could get carried away and fall in love with a Shirley MacLaine movie, a tablecloth from John's Bargain Store, or a great novel by Stendhal. It was more that side by side with her *joie,* there was a profound Levantine melancholy and perhaps a touch of the martyr—the ascetic—that surfaced most emphatically in this period.

If Dad had been a bon vivant in Cairo and continued to relish fine food and clothes, even when he couldn't afford them, Edith remained at heart the poor girl from Sakakini who went hungry when her mom couldn't put dinner on the table, except that now she *chose* to go hungry. Of course, while she was self-denying, she was exceedingly generous with us, her children. We were served, on the contrary, elaborate meals, and I, as the youngest, her fragile *pauvre Loulou,* was plied with all the stuffed grape leaves and pockets of veal I could stand. Yet even when she took me out to our favorite pizzeria on Eighteenth Avenue, she never treated herself to a slice of pizza. She merely helped herself to the crust I left behind.

My mother was such a passionate believer in the power of fasting, she tended to go beyond what was biblically mandated. If one of us became ill, she fasted for our recovery. When she had a premonition of disaster, she fasted. She fasted when she had a bad dream. She fasted when *we* had a bad dream, certain it was a thinly veiled prophecy that would come to pass in some spectacular fashion.

Ancient rabbis believed that you could reverse certain tragic outcomes by fasting. Back in Cairo when I became grievously sick with cat scratch fever, my mother fasted constantly, certain God would listen more closely, and perhaps he did: My mysterious fever vanished. She fasted when she was worried about one of us and wanted to bring about a change in our fortunes. And here in America, she dedicated several fasts to Suzette, asking God to bring her errant daughter back home—or at least see to it that she got married.

My mother's belief in the power of fasting was not uncommon in the Middle East, where faith was always tinged with mysticism. But it was utterly foreign to Jewish life in America, and in Mom's case, it also proved to be quite dangerous.

She had become very ill the previous August when she insisted on fasting in the middle of a heat wave. The doctor who was summoned to Sixty-Sixth Street seemed stunned. Here was a small, delicate woman who was underweight, seriously malnourished, and most likely anemic, who made a habit of doing without food or water at every opportunity.

And yet no one in our household—not her husband, not any of her children—had the sense to stop her?

The doctor did what he could to treat her. He ordered Mom to break the fast immediately and to avoid fasting in the future. She was heartbroken by the edict and now, a year later, there she was gearing up for the Fast of Alas, the night of Lamentations, all over again.

There was nothing any of us could say to dissuade her.

It was as if summer had been suddenly canceled. Mom was in her own way every bit as strict about religion as Dad.

The more glum the atmosphere in our house, the more I retreated to our television set. I loved *Leave It to Beaver* and *Father Knows Best* and the window they offered on suburban America, such a prosperous country of ranch homes and colonials and lawns that looked nothing like the America I knew. I had a crush on soulful, handsome Dr. Kildare that rivaled and perhaps exceeded my fondness for Maurice. I dreamed of being the perky "Bachelorette" on *The Dating Game*.

But my favorite was *Let's Make a Deal* with its perpetual cheer and bountiful prizes. If Mom believed in denying herself all earthly pleasures, Monty Hall, the host, believed in rewarding yourself with all you could possibly want—a washer and dryer, a mink coat, a La-Z-Boy armchair, a lawnmower. I fantasized about being forced to choose between door number one or curtain number two or the wads of cash Monty preferred as a sure, reliable bet.

On scorcher days, it was a struggle to stay away from the beach. Our first two summers in America, there was at least the

Loulou with César and his Egyptian friend, Raymond Benzaken, at the New York World's Fair, summer 1965.

World's Fair to keep us entertained: The rabbis had no prohibi-tions against going to the fair in this period. We traveled to Flush-ing Meadow Park again and again—sometimes only Mom and me, sometimes with my dad, occasionally with my older brother César and his friends.

We loved to wander around the fairgrounds, but from the start we had a favorite destination. We'd head to the Egypt Pavilion—home. It had a cozy snack bar called the Crocodile Café that served Egyptian specialties. One evening, we ordered *locomadis*—little balls of honeyed dough that were deep-fried yet light and savory. It was a Greek pastry that was very popular in Egypt, and Mom made them regularly. We hadn't eaten *locomadis* since coming to America, and it was as if the essence of Cairo was contained in those hot, melting honey balls, each no bigger than a grape.

We were all there huddled inside the Egypt Pavilion—even Suzette had joined us from her apartment in Queens—and we lin-gered until it was dark outside, and the World's Fair was about to close for the night, and the attendants told us nicely but insistently that we had to go.

Several members of the Shield of Young David who had been away returned for the Fast of Alas, the night of Lam-entations. Our evening began with a simple meal, which we ate silently—soup and some boiled chicken with rice on the side. Dad left to go to services down the block after turning out all the lights in the house.

I gulped down one last glass of Coca-Cola for sustenance. While Edith was undaunted by the prospect of going without food or drink for a full day and night, I was absolutely terrified. From the moment the fast began, I worried that I'd be thirsty and need a drink of water.

It was considered *haram*—a sin—strictly forbidden to have even a drop of water.

Children had dispensation, of course, but it was important for us, the girls behind the divider, to fast exactly like the adults. To be a grown-up meant you fasted and prayed. Of course, it was considerably easier on Yom Kippur when my friends Diana and Gracie and the others were there with me, and fasting became a competition to see who could last the longest. Tonight, I was going to be alone.

There was still light outside when Mom and I finished our supper and walked around the corner to the Shield of Young David. We could see other families trudging toward any one of the dozen temples in the neighborhood, including the Congregation of Love and Friendship where Dad was already ensconced for the night, and the Synagogue Without a Name next door. It was a grand old structure and the Jews who prayed there were *schlecht,* as Mom would say in Arabic. It meant that they came from Eastern Europe, worlds apart from us. We had nothing to do with them, nor they with us. They were every bit as distant and apart as our Italian Catholic neighbors.

When Edith and I entered the sanctuary, I almost didn't recognize it. The chairs had all been removed and the lights were off. The thick velvet curtain that covered the Ark had been taken down as had all the festive embroidered coverings over the lectern at the center, where Mr. Menachem usually stood and chanted the service. The synagogue was stripped bare—like a house in mourning, like a home where a loved one has died.

All around me, people were seated on the floor, including Rabbi Ruben. In the women's section, I noticed Mrs. Menachem; her hair was tucked under a simple kerchief, instead of her usual wig. She sat on the cold linoleum floor crying as she read from her small prayer book.

My mother shyly went to the back where the chairs had been stacked, carried one to the front of the women's section, and sat down. When Madame Marie arrived, she did the same. Both were deeply observant but a tad skeptical of the religious customs we were now encountering. Back in Cairo, my mother told Madame Marie, women didn't sit on the floor in synagogue. "Les Américains, toujours ils exagèrent," Mom said—Americans always go overboard.

I loved the idea of sitting on the floor.

I found myself a spot and tried to emulate the other women. But I quickly began to feel uncomfortable. The floor was so hard, and I hadn't brought a pillow with me. Though our section was more full than it had been all summer, I was still the only child. I couldn't chat with the older women—they were all lost in thought and teary eyed. I couldn't even engage in my favorite activity, peeking into the men's section—the lower part of the divider was made of solid strips of plywood, and there were no holes through which I could peer.

It was getting darker and darker inside the sanctuary when we began the critical reading of the night, reciting the Book of Lamentations, Jeremiah's ode to lost Jerusalem. It began with the single word *Alas*, which we repeated out loud again and again, like a dirge. Alas, the city that was great has become like a widow. Alas, the enemy has prevailed. Alas, those who feasted extravagantly now lie destitute in the streets. Alas, those who were brought up in scarlet clothing wallow in garbage. Alas, women and children wander about with nothing to eat and no one to comfort them. I noticed that many of the women around me were shaking and weeping.

Once we finished the Book of Lamentations, everyone rose to leave. There was none of the joking and bantering typical after a service when groups of friends would gather before going home.

People walked out subdued, their heads lowered. My mother took me by the hand; she didn't say a word to me all the way home.

The house was pitch-black when we arrived. My father had already come back and was lying down in his room. Mom made her way directly to the bedroom and went to sleep. If the point of the holiday was to relive the despair and desolation of an ancient calamity, then my parents had done a magnificent job—I felt completely bereft.

I couldn't fall asleep. I had never liked the dark and would always ask Mom to leave at least a small light on in the house, but tonight she had refused. *Haram,* she said; it's a sin. I stayed awake much of the Night of Lamentations wondering how I was going to survive the fast and how Mom would get through it. What if she needed a doctor again? I was already thirsty.

I must have dozed off at some point, because I felt my mother shaking me, saying it was morning; we had to go back to the synagogue to pray.

*F*inally, it was over. Come sundown, we went again to the Shield of Young David to hear Mr. Menachem recite the last prayer of the holiday, and everyone whom I'd seen crying the night before now appeared excited and happy. Mr. Menachem recited the Verse of Consolation—Nahamu, "be comforted"—and it was the antidote to all of our laments. Mom turned to me and repeated, "Nahamu, nahamu," as if we were going to a party.

We snuck out of temple: There was much work to be done before Dad came home. My mother seemed weak but cheerful—she had survived the fast without so much as a fainting spell. I had broken down and eaten in the late afternoon, but I was still pleased with myself—it was the longest I had ever fasted, and I felt that I was gearing up for Yom Kippur, which was only a month away.

At home, Edith retrieved several lemons from the refrigerator and began squeezing the juice into a large bowl. After adding sugar and some fresh mint leaves—she placed the lemonade in the freezer to chill, a family tradition almost as sacred as the fast itself.

Then, she took out several trays of cheese pastries she had baked the day before and placed them in the oven.

They were called *filo be gebna* and they were her specialty—strips of filo dough stuffed with shredded Gruyère, which were so light and buttery and warm that the cheese melted inside the thin papery dough. That was to be was our entire dinner—the cheese pastry and the lemonade. "Il faut manger légèrement," Mom explained—it was important to eat lightly after a fast, no matter how hungry you felt.

Only when my father came home did we finally break the fast. As we gathered at the table, Mom handed him the first glass of lemonade and then passed me a smaller glass; she served herself last. I gulped mine down and demanded another, then another. It tasted both tart and wonderfully sweet all at once. The house was ablaze with light. As I gobbled up *filo be gebna*, I realized that my sadness of the previous night—the bleak feeling that had taken hold of me as I lay there in the dark—had vanished and I was hopeful again.

· 11 ·

The Messiah Is
a Woman

I have always suspected that my family had a fatal flaw—
almost a genetic defect—dating back hundreds of years.
For all our pride and sense of majesty, our ability to rise
and distinguish ourselves, we also had a history of suffering spectacular falls. But what Mom and Alexandra blamed on the *mauvais
oeil*, the evil eye that seemed to shadow us at every turn, I attributed to our tendency to embrace false messiahs.

Literally so. In their most arrogant years, when my ancestors,
the Laniados of Aleppo, were at the height of their powers—the
rabbis and scholars who guided this ancient city—they fell under
the spell of Sabbetai Zvi, the son of a poultry dealer from Smyrna
who purported to be the Messiah. In the late 1600s, Rabbi Shlomo
Laniado used his considerable influence to galvanize all the Jews
of Aleppo.

Their savior had arrived.

Now Aleppo was a passionate community, a community of absolute believers. They had prayed so fervently and waited so long for the Messiah to come. It made perfect sense that he would choose their city to reveal himself. Rabbi Laniado headed the Rabbinical Court, and when he embraced Sabbetai as the Messiah, no one doubted his word. Any day now Jews would be gathering from all four corners of the earth for the return to Jerusalem. The revival of the dead would follow immediately.

Aleppo was in a state of ecstasy. Suddenly there were sightings of the prophet Elijah. Some spotted him praying in the central synagogue, wearing a flowing white robe, ready to usher in the Messianic Era.

Elijah may have visited seventeenth-century Aleppo, but the Messiah did not. Within a few years, Sabbetai Zvi was exposed as a fraud, an impostor, a huckster. Thrown in jail and condemned to death, he converted to Islam as a way to avoid execution; and his followers, those who had continued to hope against hope—the rabbis in my family among them—had no choice but to accept the fact that it had all been a lie. The Messiah hadn't come after all.

The Laniados suffered a shattering blow to their honor. Yet they somehow survived and the rabbinical dynasty that ultimately produced me and my family endured centuries more. The community was even prepared to forgive them. Because their faith was so overpowering, their judgment occasionally failed them.

There were altogether eleven consecutive generations of chief rabbis in my family, more than any other rabbinical Halabi dynasty. One was even crowned king of Aleppo. They continued to lead the community. They resumed waiting for the Messiah.

More than three hundred years later, I was still waiting. I was in my own way as intent and anxious for his arrival in Brooklyn in the winter of 1967 as Rev Shlomo Laniado had been in Aleppo back in 1665.

My rabbis at Hebrew school had me all fired up with their stories about the Messiah. I was constantly learning about *Moshiach* and given ever-more wondrous accounts of all he would accomplish when he finally came. He would heal the sick among us. He would reunite lost family members. He would end war, hunger, and poverty. He would abolish all sad holidays and fast days, starting with the Fast of Lamentations. He would lead us all back to Jerusalem—the living as well as the dead.

I viewed the Messiah's coming in intensely personal terms. What mattered the most to me wasn't the prospect of peace on earth but the possibility of seeing Alexandra, my maternal grandmother who had known me only briefly as an infant when she left Egypt and moved to Israel. She had died a few years later, lost and alone in a Promised Land devoid of any promise, without seeing any of us again. I wanted to be reunited with Baby Alexandra, my sister who had lived only eight days before succumbing to typhoid fever. I expected to be rid once and for all of cat scratch fever, whose return I always dreaded. And because the Great Temple in Jerusalem was to be rebuilt immediately, Mom would never again have to mourn its destruction by fasting. I believed all this and more would come to pass once the Messiah came.

I didn't think of myself as a gullible child. On the contrary, I'd question my rabbis closely, methodically, almost obsessively. How was the Messiah going to lead us to Jerusalem, I wanted to know—through what means of transportation? Boats? Planes? Would it happen overnight, or in the course of several days? How would Baby Alexandra return—as an eight-day-old infant, or as a child a couple of years older than me, the age she would have been had she lived, or as a grown woman?

At least one scenario they offered—underground tunnels that would whisk us all, the living and the dead, to one spot in the Holy Land—was thrilling to me but also unsettling. My first

Hebrew school teacher in America, Rabbi Baruch Ben Haim, a kindly, avuncular man I adored, had encouraged me to ask questions in class, and he always offered such vivid descriptions of the Messianic Era that I'd find myself thinking about the Messiah's arrival instead of concentrating on my homework or studying for a math test.

But since the start of the year, I was no longer with Rabbi Baruch. I had progressed to a more advanced class with older children and a brand-new teacher, Rabbi Saul Kassin—a disciplinarian who scared me half to death.

Rabbi Kassin belonged to a hallowed rabbinical dynasty of his own. Indeed, back in Aleppo our two families had clashed over the centuries, each challenging the other's supremacy on points of custom and Jewish law, and so perhaps it was natural that I would chafe at his air of absolute authority.

Of course, the difference was that the Kassins still wielded enormous power here in America. Rabbi Kassin's father was chief rabbi of the Syrian-Jewish community in Brooklyn, a position of honor and responsibility. The Kassin dynasty had endured while the Laniados' rabbinical lineage was alive only in people's memories, evidenced by the fact that eyes still widened at the mention of our name. I had found it useful, even in our vastly reduced circumstances, to say it loudly using the Arabic pronunciation—Leh-nee-ad-do—when I was introduced to any of the Syrian-Jewish grandees who lived in Bensonhurst or around Ocean Parkway. Though some had been in America more than half a century, they felt a deep connection to Old Aleppo and were duly impressed. We may have lost our fortune, but to this community of mine where name and family honor mattered above all, we were still the aristocrats.

I veered between my dad's Halabi arrogance and my mom's crushing humility.

Rabbi Kassin seemed oblivious to me and my pedigree, and in

his classroom there would be none of the bantering, the exchange of pleasantries and rabbinical arguments that had made Hebrew school with Rabbi Baruch so thrilling and such fun.

To begin with, boys occupied all the prime seats at the front of the classroom while girls were relegated to one row in the back. The boys were under pressure to prepare for their bar mitzvahs in a couple of years, whereas Hebrew studies for the girls were treated as an afterthought, a luxury.

My heart sank as I took my place night after night at my little desk against the wall, squeezed in next to Lillian Mosseri, a sweet, even-tempered Egyptian girl who seemed not to care much about Hebrew school or Rabbi Kassin or even the Messiah. It was hard for me to contain my dismay at how much had changed.

Rabbi Baruch had run a meritocracy where girls were the rising stars. In the course of a couple of months, the boys, so cocky and sure of themselves, seemed to fade in importance as my girl-friends and I dominated the class. We were leading a quiet revolution—our own miniature women's movement on Sixty-Seventh Street—showing girls to be equal to boys in a realm always denied to us—the realm of religion, of God.

I'd linger after class most nights, hoping to continue my exchanges with Rabbi Baruch. He loved to tell stories about the coming of the Messiah, elaborating on all the miraculous changes that would come about. I was so mesmerized I became persuaded that all of my family's recent woes—our forced departure from Egypt, the rocky transition from Cairo to Paris to New York, my parents' uncertain means—would all be behind us in a matter of weeks or months, once the Messiah came.

A drama was unfolding at home that made it imperative for the Messiah to arrive.

My brother Isaac had decided to follow Suzette's example and leave home. He was in high school, and catalogs were arriving

in the mail from faraway schools, colleges thousands of miles removed from Sixty-Sixth Street, including the Air Force Academy
in Colorado Springs.

My parents were distraught. They had lost Suzette; they were
about to lose Isaac.

The family was breaking up piece by piece, member by member,
and what could either of them do to stop it? That was America
for you—a land where children went away and parents were left
behind and family was no longer sacrosanct.

I, too, felt anxious about Isaac leaving.

"What if the Messiah comes while you are gone?" I asked him
one day.

He burst out laughing. "That is the best argument I've heard
for me to stay," he said.

But he didn't sound convinced.

My brother wasn't the only one plotting his escape. My sister,
who had continued her wanderings from one Queens high-rise
to another, now spoke of moving even farther away from us, to
Miami. In 1967, the women's movement was starting to kick into
gear with an almost messianic fervor. Its leaders offered an idealized vision of a world where women would be freed from their
lowly roles and inferior status. Instead of being prisoners of their
homes, they'd go off each morning to work in shiny offices where
they'd at last be on equal terms with men.

Salvation would come through a woman's work life, not her
home life.

Suzette had embraced the movement while it was still inchoate and took it several degrees beyond its most fervid leaders. Her
views weren't simply extreme—she tolerated no dissent. She was
utterly contemptuous of marriage and family, and she repeatedly
suggested I could do without both. I devoured her every word, the
most rapt audience she would ever have.

But I was also torn. The messages about women both reso-
nated and conflicted—perhaps fundamentally—with what Mom
had taught me about rebuilding the hearth. Edith spoke of home
and family as sacred and was profoundly sad because it was clear
that ours was unraveling.

"Loulou, il faut reconstruire le foyer," she would remind me
again and again. You must rebuild the hearth. It was as if she believed I really could put back the broken pieces of our family in a
way that she and Dad couldn't. It was the recurrent nightmare of
America, as my parents were experiencing it—a place where children left, and home, instead of being central, became meaningless.

This was more true than ever with the women's movement—
it seemed to be mostly about abandoning home and hearth. No
need for husbands or families; only realizing the self through
work mattered.

I applied the radicalizing ideas of "women's lib" to the only universe I knew, to the twin worlds of my synagogue and my Hebrew
school, to the divider I hadn't been able to breach and to the classroom where the status of girls struck me as shockingly, woefully
unequal.

One evening after Hebrew school, I headed for Bay Parkway,
whose lights and wide vistas and lively streetscapes I found
bracing and hopeful. It was very different from cozy Eighteenth
Avenue where little stores were packed together one next to the
other. Bay Parkway was more stately, its storefronts larger and
more imposing, and there were none of the cheap bargain stores
that characterized *La Eighteen* as Mom called it.

On the second floor of a storefront, a new school had opened,
offering judo and karate lessons for adults as well as children. Spectators Welcome, the sign on the door said. I made my way inside. A

class was starting, so I took a seat by the wall, trying to be as unobtrusive as possible. A burly man was lecturing to a group of children on the principles of karate. I saw boys my age in splendid white uniforms with white belts. The instructor taught them to bow to each other, and then showed them how to properly tie their belts. The instructor wore a black belt and when he and a colleague did a demonstration for the children, the two men leaped high in the air and lunged at each other and fought ferociously. Up close, karate seemed more brutal than I'd imagined it watching Emma Peel. In her black leather catsuit, she managed to be tough but supremely graceful, delivering elegant blows that made short shrift of her enemies.

Still, it was all thrilling to me. I noticed that when the two men were through, they bowed to each other and maintained a courtly demeanor.

I was suddenly sure that was what Mrs. Peel would counsel— karate lessons, not Hebrew lessons. That is what I needed.

I had tried to learn it on my own, marching off to the library for "how-to" manuals in martial arts. I took home a book with diagrams and step-by-step directions on how I could strengthen my hand, along with mesmerizing accounts of men who could smash a brick or thick pile of wood with a single chop of the hand. But when I attempted the karate exercises, none were as simple as they appeared—it hurt even to smash a thin piece of wood.

The school on Bay Parkway was a much more promising venue. I stood there watching as the instructors demonstrated more martial arts moves.

But I was afraid even to ask how much they charged for a lesson.

One Sunday, I announced to my family I was quitting Hebrew school. I'd had enough.

When Mom balked, I reminded her that I was going to be a secret agent when I grew up. I had no need for Hebrew classes or Bible instruction. Better she should send me to study judo and karate. My mother seemed taken aback by my vehemence. She didn't have strong feelings about Hebrew school—she was much more intent on my mastering French grammar and culture to prepare for the lycée. Even so, she didn't like my idea of quitting—it went against her core belief that the more I studied, the better. My father sat there impassive and removed, unable to grasp my anguish. He would happily have taught me more Hebrew, given me special lessons to make me a star again, as I had been in Rabbi Baruch's class. But that wouldn't have helped: In Rabbi Kassin's classroom, it didn't matter how much a girl knew or didn't know.

I had resolved to skip my Sunday morning class altogether and was refusing to leave the house.

César was dispatched to talk some sense into me. Whenever there was a hopeless dispute, my oldest brother would mediate. It was a role he'd had since he was a little boy. Mom revered his judgment and she loved to say, "Il faut parler a César"; we should turn to César for counsel.

I tried to make my case. I planned to become a world-class spy, I explained, and Hebrew school was a waste when I wasn't even learning much anymore. What I needed was to acquire the skills that would help me as a secret agent—judo, karate, even target practice.

My brother listened and nodded intently. He agreed that I was making a fine career choice and Emma Peel was an outstanding role model. But wouldn't Hebrew help me as I roamed the world on my espionage missions? he argued. Wouldn't I have an edge if I absorbed whatever Rabbi Kassin and the Shield of Young David Hebrew school still had to teach me, and *then* went to work as a secret agent?

As for Mrs. Peel who was so intellectual and worldly, she would certainly appreciate my learning another language, César calmly pointed out.

By the end of our chat, I had not only shelved my plan to drop out of Hebrew school but was vowing to work even harder there for the sake of my future as an international woman of intrigue.

I put my coat on, raced over to Sixty-Seventh Street, took my seat, and tried to focus on what Rabbi Kassin was saying. In the days that followed, I told myself I had made my peace with staying put—at least for now—though I wasn't entirely at ease.

A question was gnawing at me. When we'd learn about the Messiah, it was always about what "he" was going to do once "he" arrived, how "he" was going to bring about global peace, how "he" would revive the dead and rebuild Jerusalem. But it was no longer clear to me that the Messiah had to be a man. All around me were tough, formidable women, capable of great feats—women like Mrs. Peel. Was there in fact a rabbinical prohibition against one of them becoming the Messiah?

One evening as my Hebrew school class got under way, I raised my hand and waited for Rabbi Kassin to call on me. I was anxious: I knew what I was about to ask was seditious, and I had a feeling Rabbi Kassin wouldn't like it one bit. But I was determined. He walked toward the back of the classroom and nodded in my direction.

"Rabbi, why can't the Messiah be a woman?" I asked.

He looked at me ever so briefly before turning away. "Because he can't," he snapped, and that was that.

It all happened so fast I don't think my classmates noticed how crushed I was by the exchange.

When I arrived home, I climbed into bed and pulled the covers over me. I felt as crumpled and deflated as on that Saturday I had tried to breach the divider and the boys had shooed me back into the women's section. I couldn't get Rabbi Kassin's terse reply out of my mind—or the fact that he didn't even think I was entitled to a thoughtful answer.

I kept my promise to my brother and continued to attend Hebrew school. But it was as if I were on automatic pilot. I had no enthusiasm for it anymore, and I paid little attention in class. I didn't raise my hand or participate in discussions. Mostly, I whispered and giggled with Lillian and the other girls in the back row. Girls were never expected to be serious students.

I didn't think much about the Messiah anymore. I no longer expected to meet up with Alexandra again. My faith was still there, of course, deep and abiding, but it wasn't absolute anymore, it wasn't all-consuming the way it had been.

It was as if the rabbi's shattering, dismissive answer to my question was propelling me in a very different direction than the rebuke had intended, making me a riper target for assimilation. I was now even more open to all those dubious modern values that Rabbi Saul Kassin and others in that compound on Sixty-Seventh Street—downstairs at the Hebrew school, upstairs in the women's section—were trying so hard to vanquish.

· 12 ·

The Tragedy of the
Navy Blue Blazer

My mother's quixotic campaign to get me into the Lycée Français de New York seemed to be finally paying off when we received word in the spring that I would be permitted to take the school's entrance exam.

Edith was overjoyed though she warned: "Il faudra beaucoup étudier"—I'd have to hit the books and study hard—really hard.

At least I was being given a shot.

That was all she wanted, and it was all she received—no guarantee of a place for me in the fall; no invitation to come to East Seventy-Second for a grand tour of the magnificent mansion the lycée occupied or a meeting with Monsieur Galy, the school's distinguished president; no hint of a scholarship.

Yet my mother's natural optimism and ebullience took over. The future suddenly seemed dazzling with promise—Loulou off to the lycée. Loulou attending a private school. It was as if by that alone the family's lost honor would be restored and the hearth rebuilt. The Lycée Français de New York was almost as prestigious in her mind as our beloved Lycée Français de Bab-el-Louk, with its vast campus and stylish pale gray uniform and Parisian teachers. Mom had conveniently blotted out what happened at the end, how Nasser had booted all these exquisite French *institutrices* out of the country and we were left with a mediocre Egyptian faculty whose command of even basic subjects was questionable.

My mother was especially excited about the New York lycée's uniform—the dress code mandated a navy blue blazer with gold buttons, which was to her mind the epitome of style. Edith had always been in love with navy blue. It wasn't merely a color to her but an identity, an emblem of the lifestyle she so wanted me to have—elegant, refined, aristocratic, European rather than American, at least the America we knew. The lycée even had its own gold buttons with a big capital "L" in an elegant serif font that mothers were supposed to affix to the blazers, along with a crest. It was all such a dreamy prospect for Edith—the notion of sending me off to school in Manhattan every morning in a gray pleated skirt and white blouse with *une jaquette bleu marine aux boutons dorés*.

I was by no means as excited by what lay ahead. The lycée had given us a syllabus outlining what children my age were supposed to master in various subjects. There were also detailed reading lists with books in English and French. And for good measure, an administrator sent over a couple of ancient secondhand grammar books to help me prepare.

It all seemed terribly daunting.

Edith decided that we should pick up the pace of our private lessons. I had a glimpse into the formidable teacher who had been a star at L'École Cattaui and captured the heart of the pasha's wife. Each day until the entrance exam, I would have to sit with her after school (and after Hebrew studies) and learn to conjugate a verb in several tenses, or be given a *dictée* straight out of the pages of the lycée's frayed textbooks. These contained passages from great and obscure works of literature and became the basis of the spelling tests and pop quizzes Mom gave me every night. The book had an inscription from a long-ago teacher to their student. If I didn't care about you, it said, I wouldn't even bother, "mais je vous aime et je vous gronde beaucoup"—but I love you and I am constantly scolding and demanding more of you.

*I*t was the early spring of 1967, and I was now completely at ease with English. I was so pleased with myself and how I'd mastered the language that I lorded it over my siblings and parents. They all had pronounced accents they couldn't shed.

One morning, when a repairman came to the house, I airily called out, "I am the only one who speaks English." I insisted that he deal only with me. My family—César in particular—found my display of arrogance galling and months later, they were still poking fun, mimicking me saying, "I am the only one who speaks English."

But even as my command of English grew, I was losing my fluency in French. Mom spotted the danger ahead—all around us were refugee families who'd arrived a year or two earlier than we did from Cairo or Tunisia or Morocco, and the children had become so thoroughly Americanized they didn't speak a word of French anymore. She was determined this wouldn't happen to me. French, not English, was my native language, she kept reminding me—French was my culture, my true heritage.

We spoke only French at home now and instead of the radio, we'd listen to Charles Aznavour and Dalida on the small blue record player my brother had bought with one of his first paychecks. We owned a handful of records and they were all of French singers—Aznavour, our Sinatra, was my favorite; I adored his veiled, romantic voice. I could listen to him over and over singing "Donne Tes Seize Ans," a song about a young girl who is about to have her sweet sixteen, and Aznavour tells her how splendid it will be to grow up, how the world will be hers for the taking. I couldn't wait to turn sixteen, certain life would be like the Aznavour song.

Then there was Dalida; for Mom, she was in a class by herself. My mother adored the former Miss Egypt whose throaty, sentimental ballads made her cry because they conjured up the life left behind—that world, now lost, where every one of us talked in the same intensely emotional way that Dalida sang. Dalida was the hometown girl who'd made good. She was the one who had left and never left. With her extraordinary beauty and talent, she had captivated all of Europe yet never forgot her roots in Shubra, the humble neighborhood where most of the Italians in Cairo were clustered. Even when she was the toast of Paris, Dalida still came back to perform at the swell old Cinema Rivoli and people would line up for days simply to buy a ticket. And later, after the exodus, there wasn't a single Egyptian Jew in Brooklyn—or Paris or Geneva or Milan or Adelaide or Rio or Montreal—who didn't continue to worship Dalida, who didn't mourn the stacks of her albums they'd been forced to leave behind, who didn't recall her magical performances at the Rivoli when she'd declared that she was a Cairene first, last, and always.

Yet, despite Mom's best efforts, four years after leaving Egypt, I wasn't anywhere near as fluent as I'd been. I had fallen considerably behind my French peers. But I was still ahead of my classmates at PS 205.

It seemed extraordinary to me that after all these years in America—from the tail end of second grade to fifth grade—I could still coast on the little bits of knowledge gleaned from my early education in Cairo and Paris. My mother had failed in her quest to have me skip a grade, so I'd had no choice but to make the best of my situation. My teachers knew they could count on me to liven up a class discussion, and they seemed to appreciate my enthusiasm, the fact that any time they asked a question or began a class discussion, my hand would shoot up to the sky as I begged them to please, please call on me.

I felt far less of a bond with my classmates; my relations with them were cordial but not much more than that, and with a few it was decidedly cool. There had been a period when I was included in the round of birthday parties that took place over the course of the year, but the invitations were less frequent now.

My class at PS 205 was an oddly diverse group. Most of the children were Irish and Italian Catholic, but there were also sizable numbers of Eastern European Jews as well as a couple of black students. These were recent arrivals, part of an effort to desegregate the public school system, and they seemed as lost and foreign and out of place at PS 205 as I had been the first couple of years.

I was instinctively drawn to them, more than to the Jewish girls who struck me as overly confident and a bit spoiled. Some came to school wearing prominent gold *chai*s, the Hebrew letter that signifies life, and seemed oblivious to the possibility of anti-Semitism, while I lived in fear of it. I could never quite understand this public display of religion, which had been my first impression of America. The Christmas after we arrived to New York, we'd paid a visit to some Brooklyn relatives and I was struck not simply by the ubiquitous wreaths and garlands and trees the Italian families showcased outside their homes, but by the electric menorahs in the windows of almost every Jewish house.

Once we were settled on Sixty-Sixth Street, I'd tried to persuade my parents to buy an electric menorah. How I loved the soft orange light that illuminated even the darkest, wintriest New York nights. And having a public electric menorah was to my mind so American—a sign that we were full-fledged members of this strange new open society. But Edith and Leon steadfastly refused me—in their eyes religion was a low-key, intimate affair, to be observed discreetly behind closed doors, as it had been in Egypt.

The Jewish girls at PS 205 were living proof my parents were wrong. They were every bit as affirmative about their faith as the Italian girls with their little gold crosses. By showing off their *chai*s they had turned Judaism into a kind of status symbol. I had a gold pendant in the shape of a Jewish star, one of a few pieces of jewelry we had smuggled out of Egypt, but I rarely wore it and when I did, I'd tuck it under my blouse or sweater. Those rare times I had it in full view, I felt self-conscious and vulnerable.

I had no real affinity with the girls of the golden *chai*s; in spite of our shared background, I was a stranger among them. They exuded such a breezy self-confidence—in their laughs and giggles, in the way they discussed Jewish holidays or tossed about their long pampered hair.

Wendy, for example, always managed to be color coordinated when she came to school every morning, even to the point of wearing diamond-patterned maroon tights that matched precisely the diamond pattern of her maroon turtleneck sweaters. Mom would buy me standard navy tights from Woolworth's and insist they went perfectly well with whatever clothes I wore—that was the wonder of *bleu marine* she insisted. Wendy came from a family that seemed a bit more prosperous than the rest of us in our working-class neighborhood. As they flourished, parents showered their young daughters with wonderful clothes and jewelry, and I noticed how even in the course of one school year, some of my classmates were appearing in ever-more opulent wardrobes, at least by Bensonhurst standards.

I was surprised the year before when Wendy invited me to her birthday party. Edith decided to go with me; she was fearful of letting me wander anywhere around the neighborhood by myself and wouldn't even allow me to walk to school unaccompanied though PS 205 was literally around the corner. Wherever I went in New York, Mom came, too, usually holding my hand tightly as if she were afraid she'd lose me in this strange new world. We made our way to a simple but nicely furnished apartment and were promptly introduced to Wendy's mom and older sister. They struck us as so gracious and charming.

The sister was studying French and struggling to master it at school. It seemed so natural: Would Edith agree to tutor her?

My mother was delighted; having me as her lone student hadn't quite made up for the loss of her career at Cattaui. Besides, she badly needed the money in a period when my family was literally scraping together pennies to get by. She assumed that the family, who lived comfortably and dressed their daughter in the nicest clothes of any girl at PS 205, would be able to pay her well.

My mom never bothered to negotiate or even mention a price for the French lessons. She had always felt awkward about discussing money. That was business, a man's realm, and the women of our refined culture steered clear of that. Besides, she simply assumed that she would be compensated fairly by this elegant, friendly family.

There was a code of honor in the Levant that went hand in hand with the aggressive, cutthroat traditions of the *souk*. We believed in gentlemen's agreements. We put our trust in unspoken understandings. And that is what my mother, Edith, thought she had with my friend's family.

On days when I was invited to Wendy's house to play, my mother would sit in another room and dutifully go over French verbs and phrases with the older sister. Everyone seemed delighted with how the tutoring sessions were going—Wendy's mother, her sister, and, of course, Mom. The family was extraordinarily hospitable; they showered us with love and attention.

But after several of these lessons, my mother still hadn't seen a dime, and she scrounged up the nerve to question them. They seemed genuinely astonished—paying for French lessons? Why, they'd assumed that Edith was doing it as a favor to the family, to help out their older daughter.

We didn't realize that the members of this American family were creatures of entitlement. They were entitled to have friends, good grades, wonderful clothes, and gratis French lessons. Since leaving Egypt my mother and I felt we were entitled to nothing.

There was an unpleasant scene that left Edith feeling hurt and used. I gathered that Wendy's parents stood their ground and hadn't even offered some sort of token payment. While I didn't witness the confrontation, I suffered the consequences.

Never again was I invited to Wendy's house or her birthday parties, and she barely spoke to me at school. I felt oddly guilty, as if I were in the wrong to believe Mom should be paid for teaching French grammar and vocabulary to a privileged teenager.

The episode only underscored my vague, inchoate distrust of American girls. They seemed to me to lack the candor of the Levantine girls who sat with me behind the divider at the Shield of Young David every Saturday.

*T*his is how Mom and I prepped for the Lycée Français exam. Every Saturday afternoon after services, we'd set out hand in hand for "The City." Our destination was Donnell, the quirky branch of the New York Public Library that had become a kind of second home to my family. With its massive collection of foreign-language books and even a children's room with a special

French corner, it was a natural haven for us. Besides, my mother felt it offered me a far better education than I was receiving at PS 205. My mother was passionate about libraries—it gave her such pleasure to step into those quiet rooms filled with books. Our two local libraries—the spacious Mapleton branch near Eighteenth Avenue and another far smaller storefront library near Kings Highway that was a room or two—were among the only places she let me walk to alone as I grew older. She trusted libraries implicitly—they were sacred to her, holy sites.

And Donnell was the Holy of Holies.

As we journeyed to Manhattan, far from the pressures of life on Sixty-Sixth Street, I had Mom's undivided attention.

We both relished these weekly outings that took us to a corner of New York we found exotic and elegant, though parts of it were neither really. After getting out of the Forty-Ninth Street subway station, we'd walk along Seventh Avenue, passing a cluster of "adult" theaters that featured images of curvaceous women and neon signs that flashed Girls, Girls, Girls. I knew instinctively not to look too closely, and besides, the doors to these establishments were usually closed, I noticed, as Mom and I hurried past them.

I was curious but couldn't understand the fuss. I am a girl, I thought—what is the big deal?

In a ritual we both enjoyed, we strolled up Seventh Avenue past the stately white Americana Hotel. We both loved the Americana, which was outclassed in our eyes only by the Hilton down the street. Once in a while, I'd dart into the Hilton, drawn by the carpeted interior and the chic people milling about the lobby and browsing through the expensive shops. It was hard to remember that I had once felt entirely at ease and not like an outsider at all in a Hilton every bit as swank and sleek as this one: the Nile Hilton, where Dad and I would retreat nearly every afternoon to the dark bar.

Those chic, quiet blocks that led up to Donnell were a window into a gilded world we could only dream of entering. We'd walk past charming old brownstones and designer ateliers with dresses in the window that looked nothing like what I saw at Mays or Korvette's. My brother Isaac had piqued my interest in those designer workshops—he told me they could make me a dress no other woman in the world would have. That is what I wanted—an outfit that would be mine and mine alone. An outfit like the ones Emma Peel wore, that would be the envy of the women's section at the Shield of Young David and make Maurice finally fall in love with me.

Our pilgrimage always included a stop at MoMA—the gleaming Museum of Modern Art across the street from Donnell. The entrance fee was so steep we couldn't afford it any more than the Hilton or the Americana, but that didn't stop us from going into the gift shop, which was free of charge. We loved looking at the little postcard replicas of all the famous artworks upstairs that were denied us. The museum was exhibiting the photographs of Diane Arbus, and some of her haunting images were available in the gift shop. I'd stare at the black-and-white photographs of impossibly forlorn people and even at ten I felt an absolute kinship with them. They were quintessentially American, yet they looked as alienated and lost as I felt.

Once Edith had treated me to a postcard or two, we proceeded to Donnell. Mom waited as I browsed and picked out my reading for the coming week—all in French, of course. I favored a novel by Jules Verne, *Vingt Mille Lieues Sous Les Mers* (*Twenty Thousand Leagues Under the Sea*), or a volume of Tintin, the comic series about a young investigative reporter and amateur detective who battles bad guys and solves case after case as he travels the world with his faithful dog Milou. I owned exactly one Tintin, *Les Bijoux de la Castafiore* (*The Jewels of Lady Castafiore*), a gift from a family friend I relished and kept reading and rereading till I knew it by heart.

Donnell didn't stock many Tintins but when one surfaced, I'd grab it—*The Cigars of the Pharaoh, We've Walked on the Moon, The Secrets of the Unicorn.* I also loved La Comtesse de Ségur, a nineteenth-century children's author unknown in America, whose books featured little girls beset by misfortune, who somehow manage to find redemption and renewed purpose through the tragedies that befall them. Those Saturdays at Donnell, I was getting a taste of a literature that was drastically different from what I was being exposed to at PS 205, where nothing we read in class captured my imagination as much as *The Adventures of Tintin.*

Occasionally, I liked to tiptoe into the young adults section on the mezzanine level. It was forbidden territory—I faced a stern librarian who peered suspiciously at me when I wandered in, and often shooed me back to the children's room. Every once in a while, I snuck past and leafed through those teen romance novels I adored—books about lovelorn young girls who wore "formals," received "corsages," and danced the night away at proms with boys.

What is a formal? I wondered. Would I ever go to a prom and dance with a boy? Will he send me a corsage? In the same way that Donnell's children's section kept me connected with my French past, its young adult library was shaping my future, teaching me about becoming an American teenager.

Then it was Mom's turn. Together we headed to the adult foreign-language collection—Donnell's pride and joy, with shelf after shelf of French novels. As if in a spell, Edith would let go of my hand and drift off on her own. I'd see her fingering Stendhal's *The Charterhouse of Parma* or *L'Immoraliste (The Immoralist)* by Gide.

She visited the Proust section as if it were a shrine. Proust occupied the highest perch in her literary pantheon, and only Flaubert stood a chance of edging him out. She would pick up a copy of *Salammbô* and read out loud the opening line and say, as she had a thousand times before, "Loulou, did you know that Flaubert worked seven years on that one single sentence?" Then, in a worshipful tone, she would recite it: "C'était à Mégara, faubourg de Carthage, dans les jardins d'Hamilcar . . ."—"It was at Megara, a suburb of Carthage, in the gardens of Hamilcar. . . ."

I had heard her quote the line so often I knew it by heart, though I never understood why it had taken Flaubert seven years.

Mom's tastes were eclectic, and I could never predict what she would choose nor could I advise her. I was anxious to help. Was she in the mood for Zola? The Goncourt brothers? Perhaps some literary criticism by André Maurois? I'd wander around the adult collection pulling out whichever volume caught my eye or whose author sounded familiar because she'd mentioned them or had read me excerpts of their work for a *dictee*. She'd glance at it absentmindedly, then return it to the shelf shaking her head no.

We'd leave with a stash of books in our arms and walk through the same streets that filled us with such wonder, pausing once again at the Hilton and the Americana as if they were our personal stations of the cross, until we'd arrive at the seedier enclaves by West Forty-Ninth Street and we knew we had to hurry home to Brooklyn.

As I crammed for the lycée exam, PS 205 decided to offer my class a tour of the junior high school we'd be attending come fall. Seth Low was located on the edge of Bay Parkway, a considerable walk from my house. I went by bus with my classmates and joined hundreds of other fifth-graders filing into a large auditorium to hear a command performance by the Seth Low Junior High School band. The school was large, impersonal, and intensely forbidding. It wasn't at all like PS 205, which for all its problems was still intimate and manageable.

Of course, none of this mattered.

The prospect of going to Seth Low was awfully remote. I was completely caught up in Mom's fantasy about the lycée and believed I'd never be a public school student again—I was simply going through the motions.

Seth Low did have a major draw—on that morning, I was on the lookout for the one person I knew who was a student there.

It didn't take long for me to spot Maurice. He was on the stage with the rest of the band, hovering intently over his bass, the strange instrument I'd never heard him play, which was always shrouded in its brown cloth cover in the corner of the living room of his family's house on Seventieth Street. The band struck up the theme to *Man of La Mancha;* the show was so popular in the spring of 1967 that everyone seemed to be humming or singing "The Impossible Dream." Maurice seemed to be focusing intently as the song came to its euphoric climax, "This is my quest, to follow that star, no matter how hopeless, no matter how far. . . ."

We were told by our hosts that each and every one of us could learn to play a musical instrument. We, too, could join the Seth Low Junior High School band. Edith didn't seem impressed by this claim or by the rest of my description of my day at Seth Low.

She had no intention, she reminded me, of letting me attend *une autre terrible école publique,* yet another awful public school.

As the day of the exam neared, Mom's lessons intensified. I practiced conjugating a jumble of verbs in a dozen tenses, one more bewildering than the next. "Loulou, le passé composé," Mom would call out, the past tense. "Et maintenant, l'imparfait et le passé simple," she'd say, switching to the past imperfect and the past historic. There was also the *passé antérieur*—the past anterior—and the *conditionnel passé,* the past conditional, and I couldn't tell one past tense from the other.

Nothing stuck.

No matter how often she reviewed the tenses with me, I was always, always getting them confused.

My mother soldiered on; she wanted to leave nothing to chance. I had to submit to constant *dictées* from the old grammar textbook. Some of the passages were so lyrical, excerpts from long forgotten works of literature, and they often had a moral lesson. I preferred the *dictées* to conjugating verbs or taking one of her impromptu quizzes. I even had a favorite: a passage called "Pierre et Lucette dans un Jardin Abandonné" ("Pierre and Lucette in an Abandoned Garden"). The ancient grammar book we used had a drawing of my mischievous namesake playing hide-and-seek with her friend Pierre far from the eyes of prying adults.

There were stepped-up math lessons—multiplication tables, complex divisions, percentages, even a hint of beginner's algebra. The quizzes she liked to give both at home and on the street became both more frequent and rigorous.

Edith was a born grammarian, a natural speller, an elegant stylist. It all came so easy to her, and I think she was genuinely puzzled by how much I was struggling, how hard it was for me to grasp these elemental lessons.

My schoolwork as well as my Hebrew classes fell by the wayside those final weeks. When Edith saw me stumbling over the *passé composé*, she'd remind me of all the pleasures that awaited me once I entered the lycée—the lifestyle that I would enjoy among "des jeunes filles de bonne familles," those upper-crust young girls from good families who crowded the lycée's halls and classrooms in navy blue blazers over exquisitely pleated skirts.

She begged me to concentrate, to please apply myself, to focus—"Loulou, tu est la rose de la famille," she'd say; Loulou, you are the rose of the family. It was all within my grasp if only I put my mind to it.

I wasn't so sure: The more I studied, the more I realized the immensity of the challenge—and the less confident I felt.

A couple of days before the lycée's exam we took a break. Mom decided that I needed a special outfit for the occasion, which she assumed would include meetings with the lycée's teachers and administrators. I was always game for a shopping expedition, although this one seemed to be more pressured than our usual meanderings through bargain stores. We agreed that Brooklyn wouldn't do—neither Mays nor our beloved Berta Bargain Store on Eighteenth Avenue could possibly have what I needed to make a good impression at the lycée. Instead, we traveled to Alexander's, the discount Manhattan department store that was slightly more upscale than the shops we patronized. It also had the advantage of being on the East Side, which gave us a hopeful feeling, as if we were inching closer to that world we were trying to penetrate.

It was a season of fanciful clothes in vivid colors. The mod look had spread far beyond London, and there were minidresses and flowing dresses with accordion pleats, and empire dresses and sheaths. I pounced on a blue-green print dress made out of delicate voile. It had a low-waisted skirt that swirled round and round, with a velvet ribbon as a belt. It was the kind of dress I could wear to a party, but Edith wasn't sure it was sober enough for the lycée, and we continued our hunt.

Finally, we spotted it—"un tailleur bleu marine," my mother squealed—a navy blue suit. It had an A-line skirt, simple and severe except for the red-and-white lace trim around the jacket and jaunty bell sleeves. Without even blinking at the price, she bought it for me.

The day of the exam, I put on my new navy suit and went to PS 205 as usual in the morning. I felt elegant and transformed and utterly superior to my classmates—as if I had already left them and Brooklyn behind for the greater empire of Manhattan. Mom came to get me at lunch; she told the principal that I had an important doctor's appointment and needed to be excused from class. We traveled together to East Seventy-Second Street, the vast luxuriant street where the lycée was located.

Once we arrived, there were no social niceties. I wasn't taken in to meet *madame la directrice* or *monsieur le président*. It was all strictly business as I was whisked to a classroom upstairs to take the test.

I don't think anyone even noticed my impeccable blue suit.

Only a handful of us were taking the exam. I was struck by a young boy my age, perhaps a bit older, who wore a formidable blue blazer with a white shirt. He seemed confident and relaxed and patrician, as if this test was merely an amusing formality that didn't faze him a bit. He chatted amiably with the teacher in a fluent, upper-class French she seemed to appreciate. I had the feeling he was a shoo-in, either because he was qualified, or rich, or most likely both.

I glanced at the test and panicked. I could barely understand a single question—not in grammar, not in composition, nor in mathematics. Nothing in the exam made any sense to me. And when the teacher gave us a *dictée*, the words sounded strange and unfamiliar and I struggled to keep up and jot down what I could. Whatever Mom had tried to teach me all these many months—all these many years—hadn't penetrated or hadn't been nearly enough. I felt as if I were hearing a foreign language.

And maybe I was.

It dawned on me that I was no longer at ease in French, my native tongue.

I am not sure how long the exam lasted. The afternoon was a blur from start to finish—the questions, my answers, what the teacher-supervisor gave by way of instructions, the charming boy's light running banter, and the fact that I kept obsessively staring at his navy blazer when I should have forced myself to calm down and figure out which questions I could tackle. My throat felt parched, but I was so terrified I couldn't muster the courage to ask permission to get a drink of water. Even in the examination room, I realized that my mother's quest was hopeless, that her graceful letters to the honorable Edouard Morot-Sir and Monsieur Galy, the lycée's esteemed president, had all been for naught.

When I went downstairs, I found Edith at the same spot where I'd left her. She had sat there, afraid to move, for the entire duration of the exam. I couldn't even look at her. As we rode home on the subway, she anxiously pressed me on the types of questions I'd fielded: Was there a *dictée*? Had I been asked to conjugate many verbs? Did I remember my *passé composé*? I didn't have the heart to tell her how poorly I had done—that I didn't think I had been able to answer more than one or two of the lycée's questions.

She wouldn't have believed me.

I was the last crucible of all her outlandish hopes and ambitions, her last chance at a new and better life. She was so convinced I had it in me to realize her dreams—*Loulou, il faut reconstruire le foyer,* she would say over and over again; you must rebuild the hearth. She was sure I would find a way to recapture my family's lost place in the world and the Lycée Français was the crucial first step.

And I had listened to her, my tender false messiah of a mother, who always seemed to be pinning her hopes—and mine—on the unattainable, who was always dreaming the impossible dream.

I imagine that the lycée found an exquisitely polite yet firm way to tell her that I wasn't going to be admitted to the school. My mother spared me the news. The lycée was simply never discussed again, and I braced myself to enter, come fall, Seth Low Junior High.

13

The Advance of the Little Porcelain Dolls

Edith finally—finally—got her wish: I was offered a scholarship to start high school in a tony private girl's academy.

The chance to escape the public school system couldn't have come fast enough for Mom or me. The last couple of years at Seth Low Junior High had been a nightmare of unwieldy classes and insipid courses, punctuated by two teachers' strikes including one in the fall of 1968 that went on for months.

That was the last straw.

Mom found it incomprehensible that teachers would walk out on their jobs. What about the children, she kept asking. It was yet another example of the fatuousness of a system she had despised from the time I'd entered it.

I had my own grievances—the ethnic divides I'd first spotted at PS 205 were even more in evidence at Seth Low, where

Jewish kids dominated the "SP" or honor class, while Italians seemed to be relegated to less intensive programs. I was transferred into the honors group shortly after I arrived, and immediately regretted it. I had long ago found a way to coexist with the Italian children who admired my academic zeal and seemed to cheer me on in my unrelenting drive to excel. Not so with the "gifted," overwhelmingly Ashkenazi kids in the "SP" class. They were clannish, animated by a sense of their own superiority over the rest of Seth Low, and didn't see me at all as one of their own. I was, to them, an outsider, an intruder. How dare I challenge their intellectual dominance?

Which, of course, I relished doing.

The girls with the golden *chai*s, with names like Gross and Grossman and Friedlander, were out in full force, chatting excitedly about their upcoming bat mitzvahs—the coming-of-age ceremony for girls that was such a distinctly American phenomenon, completely unknown to me and my friends at the Shield of Young David. The girls of Aleppo and Cairo simply didn't have bat mitzvahs.

My new classmates seemed more affluent. They dressed a cut above the rest of the Seth Low girls, in a kind of uniform that included expensive beige woolen coats. They'd come back from winter holidays with tans, talking about family vacations in Miami Beach. I had absolutely nothing in common with them, with the exception of one gentle, brainy girl named Carin Roth, who seemed to be respected by her peers, and yet devoid of their arrogance. I bonded with her but kept my distance from most of the others and wandered through Seth Low as if in a bad dream.

Then the chance to begin high school at the Berkeley Institute came along with financial aid, and Mom and I seized it. Berkeley wasn't the lycée, and it wasn't in Manhattan. It didn't even have a uniform. But it was an old-line private school, quite aristocratic in its own way—catering to venerable Brooklyn families as well as the gilded nouveaux riches.

Instead of seamstresses and firefighters and sanitation workers, I found myself starting high school side by side with the children of doctors and lawyers and even the Brooklyn district attorney, Eugene Gold, whose brainy, no-nonsense daughter Wendy became my friend.

Mom was overjoyed; "Monsieur Le District Attorney," she dubbed him.

She was so thrilled when the DA's black chauffeur-driven Cadillac pulled up to our door on Sixty-Fifth Street. The DA would be in the back with his wife—the car was so roomy it fit several of us comfortably. My friend was as nonchalant about her means of transportation as she was about her status as the district attorney's daughter. She came to school by limo every morning with her bodyguard-driver and it would have astonished her to learn that my ride in the gleaming Cadillac—even more than the scholarship to Berkeley—was the talk of my family, one of the high points of our five years in America.

The neighbors noticed, of course, as did our new landlords. In a year of changes, we had been forced to leave our beloved apartment on Sixty-Sixth Street after a dispute with the new Italian landlords. We now lived on the bottom floor of a two-family house situated on Sixty-Fifth Street, only a block from our old digs yet it felt a thousand miles away. Our apartment was much larger, but there were so few of us left we felt lost in it.

The Shield of Young David had also closed—suddenly and much to our chagrin. For months, there'd been rumors that it might shut down, but it was impossible to get a straight story as to why—what had gone wrong?

Perhaps those of us who'd remained loyal didn't want to face the simple truth—that we were yesterday's news. Our community was moving away, one family after another, and reconstituting itself by Ocean Parkway.

I was about to turn thirteen in the fall of 1969, tall and confident in the chunky brown heels and green tweed minidress I'd purchased both to enter high school and to wear on the High Holidays.

In one last-ditch effort to keep our congregation together, we gathered at a nearby Veteran's Hall for the Jewish New Year. An effort had been made to replicate our old shul down to the women's section, but I felt disoriented when I walked in.

Then I realized that everyone else was also trying to make sense of the new environs.

We couldn't quite figure out what happened, why this Veteran's Hall was all Rabbi Ruben could muster. It was packed for the New Year prayers—all the old faithfuls were there, which made me wonder why we had shut down in the first place. Surely, it wasn't for lack of attendance—Ocean Parkway or not, we still had enough Bensonhurst stragglers to fill a synagogue.

To add to the surreal quality of the day, I found myself drawn into one final melee with my childhood nemesis, Mrs. Menachem.

It was a balmy September morning, and my friends and I were chatting amiably outside on Twentieth Avenue. We returned to the makeshift sanctuary only when the Torah scrolls were brought around, a high point of the service. As the men solemnly carried the scrolls past the women's section, I reached out to kiss one, putting my hand through the new divider, which was a lot flimsier and more porous than the one we'd had at the Shield of Young David.

Suddenly, I heard a scream.

"What are you doing?" Mrs. Menachem was yelling in a voice the entire women's section could hear above the hymns and chants. I turned around to face her, thoroughly confused, not realizing at first that her words were aimed squarely at me. "What are you doing?" she repeated. She was now standing very close to me and her face was red, and I felt as if she were about to hit me.

Didn't I know that I wasn't allowed to touch the holy scrolls? she asked. That I'd reached an age when I couldn't go near them?

I was too shocked to reply; all the women were staring at us, as were many of the men. I ran out and went home. What had seemed so exciting in the morning—wearing high heels and a short dress and feeling grown-up—Mrs. Menachem had used as a battering ram to put me in my place. Once a young girl reached adolescence, she was considered "impure" according to Orthodox Jewish dogma, and by that definition, she wasn't allowed near any holy object. That was the point of the divider. That had always been the point, but while I could ignore it all these years, it was inescapable now.

I was almost a teenager, and that meant I was impure.

But I'd never thought my Levantine congregation took the law literally. The women behind the divider were always rushing to kiss the Torah, to physically grasp its maroon velvet covering as it came around. Sure enough, I heard that in the afternoon several of them had confronted Mrs. Menachem and told her she was in the wrong. How could she be so cruel—and on such a holy day—they'd said, springing to my defense.

I was moved by their efforts, but the damage was done. Mrs. Menachem had completely humiliated me—exactly as she had so many years earlier when she'd called me "a silly, silly little girl who was trying to change the world."

I was starting a new life at Berkeley; why bother with this little VA hall pretending to be a synagogue, anyway?

*I*n all the years Mom had plotted and schemed to have me attend a private school, she hadn't anticipated the daunting logistics—the fees that even scholarship students were required to pay, the wardrobe needed to keep up with my well-heeled classmates, and especially the long, arduous commute. Both PS 205 and Seth Low Junior High had been within walking distance, so while my mother worried about me constantly—worried about the quality of education I was receiving, the "ruffians" I was befriending—at least I was close by.

Berkeley, located near Prospect Park, was practically at the other end of Brooklyn. I needed to take two trains to get there every morning, followed by a brisk walk through Park Slope, a neighborhood that like so much of New York in 1969 was a strange mixture of the elegant and the seedy, with magnificent old brownstones next to boarded-up homes and dubious characters lurking on street corners up and down Seventh Avenue.

It was thoroughly nerve-racking, this commute.

Edith at first insisted on taking me to school each morning, which she hadn't done in years. She wouldn't let go of my hand until we arrived at the door of 181 Lincoln Place, the small faded brick building that housed Berkeley. I balked, anxious about what my classmates would say, and begged her to let me proceed on my own to the school door. "Mais j'ai treize ans—je suis grande maintenant," I told her indignantly. There she was clutching my hand and treating me as if I were still a child, when I was old enough to take care of myself.

I could tell she wasn't convinced; what I couldn't predict was her dramatic course of action.

Shortly after I started Berkeley, Mom announced to us that she had found herself a job: She was going to be working at Grand Army Plaza, the flagship branch of the Brooklyn Public Library.

We were all stunned—each and every one of us, including Dad, who wasn't even consulted. We had no idea she had even applied for a position, let alone said yes. One day after dropping me off at

Berkeley, she had trudged across the expanse of Prospect Park West and interviewed for a job at the imposing library with the big bronze entranceway. And though she didn't have any of the classic credentials—a college degree or even a high school equivalency diploma—she was able to draw on her vast store of knowledge and her literary sensibility to persuade the library to hire her practically on the spot.

My father was upset, but powerless to stop her; he couldn't even voice any serious objections. These days, Leon seemed so much weaker and less consequential to the running of our household. He was completely dependent on César, who gave him small sums from his earnings that Dad would then dole out to Mom. We were still struggling financially and Edith had to beg for every dollar to manage, so how could he possibly object to the idea of more cash for the family?

And then there was me. *Loulou, cette petite diable,* Mom would say fondly; "Loulou, that little devil." Now that I was attending a fancy school, I kept pressuring her for expensive clothes as well as more pocket money so I could keep up with my chic classmates.

It made eminent sense, her going back to work. Yet my father couldn't make his peace with it. Edith was changing before his eyes, and it was as if America had delivered one injury after another.

A year or so earlier Mom had gone to be fitted for new dentures; the first set, obtained with the help of a social worker shortly after our arrival in New York, had been so painful and ill-fitting that she simply never wore them. But the new ones, made at a free clinic in Brooklyn, were more comfortable and she forced herself to get used to them. She looked so much more attractive: She could flash her lovely Ava Gardner smile again.

With a few tweaks to her wardrobe, thirty years after her post at L'École Cattaui, my mother was ready to return to work at a library.

It was a part-time job and she was only a clerk. Her pay was a pittance—barely above minimum wage. But no matter, to her mind she was going back to those halcyon days working with the pasha's wife. She would be her own woman again, and more important still, she would be surrounded by books.

She was so devout, she viewed this in near-mystical terms—as if God were giving her another chance. And, of course, since Grand Army Plaza was close to Berkeley, she would be able to accompany me every morning. I hadn't banked on her indomitable will.

We reached a compromise. We'd set out together in the morning, first by subway to DeKalb Avenue, then changing trains to Seventh Avenue and the long dark station that led out to Park Slope, an area that veered between faded elegance and decrepitude. Mom would walk me to the corner of Lincoln Place and, re-

Edith surrounded by books at her desk at the Brooklyn Public Library, Grand Army Plaza, circa 1970.

assured that I was only a block or so from school, she'd finally let go of my hand and begin the uphill march by herself to the library.

It was an enormous trek, and I would find myself worrying about her as she sprinted toward Grand Army Plaza. She had a choice of routes, but none was especially pleasant; and they all required her to cross the biggest, most dangerous traffic circle in New York. Grand Army Plaza was a maze of converging streets and dizzying thoroughfares—Flatbush Avenue, Prospect Park West, Eastern Parkway, Vanderbilt Avenue, Plaza Street, Union Street—and crossing it meant taking your life into your hands.

The walk was arduous even in the best of circumstances. Occasionally, she'd cut through a small park that led directly to the archway at Grand Army Plaza, but it, too, had its drawbacks. One day, she was accosted and knocked to the ground by one of the young thugs who lay claim to this deserted patch of grass. She learned to simply run, run through it to get to work.

Once she made it to the stately old building shaped like an open book, she could breathe again. She loved the grand entrance with its shiny bronze statuettes of literary icons—Tom Sawyer, Rip Van Winkle, Hester Prynne, even Moby Dick and the Raven.

From the start, she found the library embracing and safe, a world unto itself, a family even as ours was crumbling. Suzette had taken several trips to Miami and was hoping to settle there, despite Mom's entreaties that she stay close to home. Isaac had left us a couple of years earlier, first to attend a state university in Memphis, then transferring to Bowdoin, at that time a Waspy, preppy men's college nearly four hundred miles away in Maine. Every once in a while César spoke of moving out as well. He was barely around, always off with some girlfriend or attending night school or working late.

That left Dad and Mom and me in our vast new apartment, hanging on to each other for dear life.

Ed Kozdrajski, Edith's colleague at the Brooklyn Public Library's catalog department; she referred to him as "le gentil petit Ed," sweet little Ed.

The library offered Edith a home that seemed far more hopeful. She was brimming over with stories about her colorful colleagues. She couldn't wait to tell me about the mysterious Ukrainian exile and former lawyer, Alex Sokolyszyn, whom she always addressed by the honorific "Docteur Alexis"; he was her favorite. He had a Ph.D.—it was unclear in what field—and he, too, had suffered a fall in the move to America and was forced to reinvent himself and start again. Offered a job as a cataloger at Grand Army Plaza, he acquired a reputation as the ultimate authority on foreign-language books. Erudite, courtly, and thoroughly old-world, Mom idolized Docteur Alexis.

She found another kindred soul in Fawzia El-Araby, a cataloger from Cairo. The two were constantly chatting and joking, switching back and forth between Arabic and English. When Fawzia brought a Middle Eastern lunch—falafel in pita, which

Edith favored above all other foods—she'd share it with Mom. Fawzia was a Muslim, and their friendship reminded Mom of the old days in Egypt, when her social circle had included Muslims and Christians and Jews, and those differences simply didn't matter. There was also tender Jean Nason who was handicapped and struggled to do her work despite a severe case of scoliosis, and she had trouble speaking. Mom found her so moving; Jean, she'd tell me, managed to be a superb cataloger despite her physical impairments. She also developed a bond with Ed Kozdrajski, a handsome young catalog trainee with a serious manner about him. He wore thick black-rimmed glasses that made him look stern, when he was actually very gentle and especially solicitous of her. Ed was of Polish descent but born here in New York, and he and

Alex Sokolyszyn—Docteur Alexis, as Edith liked to call him—was a Ukrainian exile who reinvented himself as a cataloger of foreign-language books at the Brooklyn Public Library.

the other Americans were in the minority. Her closest companion was Mudit Strasdins, a honey-blond refugee from Latvia by way of Canada whom Mom called Madame Mudit.

She felt at home among these exiles—expatriates and émigrés and loners every bit as lost and diminished as she was in America. They banded together in a crowded corner of the third floor of Grand Army Plaza and there, amid carts bulging with books in every language on earth and rickety steel shelves crammed with the thick reference volumes of the Library of Congress and the stacks of beige catalog cards that Mom and other clerks typed up throughout the day, they rebuilt the hearth.

There were constant parties and *café klatsches* and get-togethers. Any occasion—a birthday, a wedding, an anniversary, a holiday— called for a cake and cookies and steaming cups of coffee or my mother's favorite, *chocolat chaud*, which she believed offered her all the nutrients she needed to survive.

Docteur Alexis, Madame Mudit, *pauvre* Jean, La Fawziah, *le gentil petit* Ed Kozdrajski, "sweet little Ed Kozdrajski," who she always called by his first and last name: Her colleagues were her favorite topic of conversation. Through her telling they emerged as larger than life, major characters in the novels Mom was always beginning in her mind and could never quite put on paper save for some random jottings here and there.

Then there was the paycheck. For the first time in years, she wasn't dependent on my father or César to live. Mom now had a little money to spare, a little money to spoil me, more money than she'd had since the years with the pasha's wife because for all her married life, even the period of relative ease and comfort in Egypt, she had been at Dad's mercy, relying on his beneficence—or lack of it—to make ends meet.

The library had changed all of that—it had restored her sense of self, it had made her independent. Edith's transformation from

Levantine housewife into career woman extraordinaire was in keeping with the changes sweeping America, and at breakneck speed. The feminist movement had gone from an imploring whimper to a raucous and angry call to arms and Mom, the delicate porcelain doll who for years couldn't stand up to my father, who consistently sacrificed herself to his needs and ours, now had that most wondrous of possessions—a disposable income.

As the 1960s came to a close, my mother emerged as a small, unlikely, thoroughly unheralded emblem of women's liberation.

*M*y life had also changed. Berkeley was a cloister—a secluded and largely female universe where the outside world didn't seem all that relevant. My entire freshman class consisted of twenty girls

In the years before I arrived, my classmates—who had been attending Berkeley since they were small children—learned how to curtsy and the proper etiquette for serving tea. The school emphasized quiet and decorum—in its hallways and careworn library and wide staircase with its gleaming old wooden banisters.

Even gym was different—gentler, more graceful than I'd known it.

I wore a stylish dark green tunic over a white shirt and studied fencing, archery, and badminton—sports where I actually showed promise. Fencing, in particular, reminded me of Emma Peel. I never thought about her much these days; my obsession with her had faded with the passing of my childhood. But donning the white protective garb, steel foil in hand, took me back to those years when I'd loved to watch her thrust and parry and prayed I would grow up as fearless and talented as she.

I bonded with my teachers in a way I never could at Seth Low; Berkeley was simply a different world. I had a crush on Mr. Gardine, the elegant southern gentleman who taught us English and came to class dressed in exquisite blazers, silk ties, and two-toned shoes. I found him deeply exotic—his accent, his manner, and his assignments, which were always somewhat out of the ordinary.

But he was often stern with me; his notes to my mother warned that I was "solipsistic" and measured life "by the yardstick 'I.'"

One day, instead of giving our class one book to read and discuss, he selected a different novel for each of us—works he thought we would enjoy because they suited our individual temperaments. "*The Death of the Heart,* by Elizabeth Bowen," he said, pointing to me.

It was about a young girl of about my age who is always an outsider, who never feels at home anywhere.

Islay Benson, the aging spinster who ran the theater department, was perhaps my favorite among all the teachers. With her British

Loulou, the year she attended the Berkeley Institute, Brooklyn, 1969.

accent, 1940s hairdo, and severe clothes out of another era, she was a definite eccentric, an oddball in this fashion-conscious Brooklyn girls' school. Miss Benson favored shirts with big pointed collars and long pencil skirts that brought to mind Rosalind Russell in *His Girl Friday*. She had studied drama with Frances Robinson-Duff, "the acting coach to the stars" who had taught Katharine Hepburn and Clark Gable. A quarter of a century earlier, in 1944, she'd had a starring role in a Broadway thriller, *Hand in Glove*, directed by the same man who discovered Boris Karloff and cast him as Frankenstein. Her theatrical past could have made her haughty or superior, but I found her, on the contrary, accessible and kind; and I felt more energized by her drama workshop than by my other courses.

We practiced *Twelve Angry Women*, an adaptation of the Reginald Rose play, and I was asked to read a small part, playing one of the minor jurors. I hadn't acted since my long-ago stint as Haman, but I still had a passion for the spotlight—I loved showing off, reciting out loud, being on a stage.

I was already a seasoned actress in my own way. I was still pretending to be French—a little Parisian girl, not a Cairene. It was a role I'd assumed at Mom's suggestion in our early years in New York, when she thought it would be hurtful to say I was Egyptian. It was easy since I spoke with a heavy French accent then, and I was said to "look French."

Even now, in high school, I was still pretending. I didn't tell my Berkeley classmates very much about my Cairo origins. Nobody needs to know, I thought, recalling my mother's warnings that Americans simply wouldn't understand our background, that they'd have no sense what it meant to be from Egypt. Americans had such loopy notions of Egypt, she'd say. She worried that they would dismiss me as coming from a backward and primitive country.

Besides, being French carried so much more cachet at Berkeley.

Freshman class at the Berkeley Institute, 1969/70; Loulou, standing a bit apart from her classmates, is in the front row on the left.

The first jarring note in high school came at assembly one day, when a senior, a daughter of a rabbi, recalled going to Woodstock in the summer. I heard about flower children and the hippie movement for the first time in the auditorium of the Berkeley Institute in the fall of 1969. But that was the exception: Our lives, complete with Latin classes and dress codes, had nothing in common with the furious goings-on outside. Our teachers were clean-cut and conservative in the extreme, and life was staid—not at all about the abandon and recklessness that Woodstock and the sixties had come to emblemize.

Mom noticed a change in my friends—they were so polite, these Berkeley girls. She simply melted when Wendy Gold called and asked her in French, and with such refined manners, whether

I was home: "Est-ce-que Lucette est la?" That was exactly the sort of friend she hoped I would have.

When Berkeley announced its first dance of the year—an old-fashioned "mixer" with the boys of Poly Prep Country Day School, an elite Brooklyn private school—there were so many rules and restrictions in place that having fun seemed almost beside the point. First and foremost, a strict dress code was put in place. Dresses and skirts had to be of a certain dignified length, and pants were off-limits. Jeans, of course, were thoroughly out-of-bounds. Some of my more fashion-conscious classmates were dismayed; they argued for permission to at least wear the sleek pantsuits that were all the rage.

Weeks of deliberation and planning ensued. The dance was going to be chaperoned, our parents were assured. There was to be no alcohol, and drugs were quite unknown. The mother of my classmate Kim had volunteered to drive both of us to Berkeley and planned to take me home later that night; with that guarantee, off I went to my first dance.

Yet for all the build-up and worry, it was an awfully tame affair. I walked upstairs to the assembly hall, which had been decorated and transformed into an elegant ballroom. There were the boys of Poly Prep in jackets and ties, eyeing us as we came in. Several of my classmates had dressed up as if they were going to a wedding; one student wore a striking, formal red chiffon dress. I wore a pale pink brocade sheath I had purchased the year before at Stern's, the grand old Manhattan department store, as it was going out of business. I had worn it only once, to the Newark wedding of one of César's boyhood friends. When I'd gone out on the dance floor, the band struck up, "You Must Have Been a Beautiful Baby," as the conductor smiled and waved at me. I never felt quite as wonderful or grown-up as that night in Newark, and I figured it would bring me good luck to wear it again. It fell chastely past my knee so I easily passed muster with Berkeley's dress code police.

An earnest, quiet boy about my height came over to introduce himself and ask me to dance. His name was Henry Finkelstein, and I was too taken aback to say yes or no—to say much at all— but I spent the evening only with him, dancing and listening to him chat about "Poly" as he airily called his school. He struck me as sweet and unassuming—in no way a snob. I noticed that he had wonderful green eyes—Maurice's eyes.

He phoned a few days later: Did I want to go to a dance at Poly?

Henry came from an entirely different background—a different world really—than mine. His father was a surgeon, and his family lived in a brownstone in Park Slope, by Prospect Park. His mom volunteered at the Brooklyn Museum, and his two younger sisters were also enrolled at Berkeley. One, Susan, was only a grade below mine. She seemed to be eyeing me curiously in the days after the dance, and I wondered what on earth her brother could have told her about me.

"Loulou, tu va sortir avec un garçon?" my mother asked me, sounding an awful lot like Dad in his stern Old Syrian mode—Are you planning to go out with a boy? I argued that a dance at Poly Prep wasn't really a date—though, of course, that is exactly what it was.

Henry was *un garçon de bonne famille*, a boy from a good family: I figured that Mom would like that. But my efforts to soothe and reassure her failed. I couldn't appeal to my father, of course, who would have been even more intransigent. Strangely, he didn't even figure into these discussions. He was so withdrawn that the task of bringing me up as I edged into my teenage years was left entirely to Mom and César.

Now pushing seventy, Dad was tired, defeated from the skir-

mishes and all-out battles with my older sister. He seemed loath
to get involved with raising another teenage daughter, so César
stepped into the breach and began acting as my de facto father.

My mother found a kindred soul, someone in whom she could
confide her worries about me and my schoolwork, my friends, and
now, my emerging romantic life. The two conferred at length and
came back with the firm answer: no.

It didn't matter that Henry's father was going to be driving us.
The answer was still no.

Finally, as the entire evening risked being torpedoed, Henry's
mother personally intervened. She promised that she would accompany us; both she and her husband would take us to the school
dance, she assured Mom. Mrs. Finkelstein was so gracious and
persuasive that Edith found herself saying yes even as I contem-
plated the prospect of going out with a boy and meeting his father
and his mother all on the same night. I had never gone out with a
boy before. The years that I had spent dreaming of Maurice, plotting ways to earn his love, had resulted in . . . nothing.

Yet here I had gone off to a dance and ended up being asked on
a date after one evening.

Growing up certainly had its advantages.

In the days prior to the dance, I walked up and down Eighteenth Avenue, anxiously combing the stores with Edith in tow.
Somehow, the bargain shops I'd always found so abundant seemed
inadequate for the occasion, and I wandered restlessly from one to
the other. I had no idea how to dress for a first date. Provocatively?
Demurely? The Saturday of the dance, I finally settled on a pale
blue flowered dress that was rather prim. It had long sleeves and
a small white collar and it grazed my knee so that I looked the
picture of modesty.

That evening, as I put the final touches on my outfit, I could
feel my mother and brother inspecting me. Edith looked a bit wist-

ful. I probably hadn't seemed grown up to her until that night, when I left to go to the dance with Henry Finkelstein, both his parents waiting for us as promised in the car downstairs, even as his dad curiously eyed our simple two-family house.

After I left, Mom sat down and penned a letter to my sister.

Dear Suzette:

One hour ago, they came to pick her up—this little young man, as tall as Loulou, who is now taller than me, his mother, so beautiful, so chic, and his father, too—he's a doctor in Manhattan with the most stunning car imaginable. Loulou, what a little devil. . . . With each invitation she receives, César and I have to come up with a new wardrobe for her.

Henry Finkelstein and I didn't dance much that night. Mostly we wandered around the grounds of Poly Prep, a vast campus of ponds and rolling brooks and wooded enclaves in between stately old buildings. It was hard to believe we were in Brooklyn. There was a dreamlike quality to our walk, and I had trouble focusing on what Henry was saying. He was clearly overjoyed to be showing off his school; the campus was more than twenty-five acres, he boasted.

Finally we reached one of his favorite buildings—the gym where he said he was studying wrestling. I must have looked surprised—he was so slight, not much taller or heavier than I was. It was a special kind of wrestling class Poly offered, "For ninety-eight-pound weaklings," he volunteered. We walked round and round a large room in the basement that had mats spread out all over the floor; on the wall were vintage black-and-white pictures of student wrestlers from years gone by. I peered at the pictures intently, as if seeing them was the point of the evening. I wasn't really sure what to do.

The dance over, his parents drove me home and that was that.

It had been a perfectly amiable evening, but to my secret relief, I never heard from Henry Finkelstein again. It would have been agonizing to ask Mom if I could see him again, to seek César's permission.

And that is how my first relationship with a boy ended—before it had even begun.

We weren't prepared for the perils of a Park Slope winter, which somehow felt colder than cosseted Bensonhurst. Neither Mom nor I knew how to dress properly for our daily expeditions to Lincoln Place and Grand Army Plaza—the glacial subway stations, the walk through wide-open thoroughfares buffeted by wind. I, at least, had a knitted scarf and a hat. But Edith insisted on wearing the same little nylon kerchief she used in synagogue to cover her head, and that she'd purchased for ten or twenty cents from Woolworth's. She had a collection of these kerchiefs in a drawer at home, and on extremely cold days, she simply wrapped another, identical one around her neck and ventured outside; I wondered how those two flimsy swatches of fabric could offer her any protection.

She had recently bought a long woolen coat that came down to her ankles. It was too big, more bulky than warm really, but it was of that dreamy shade of blue she fancied above all other colors, *bleu royale,* with gold buttons. After she kissed me good-bye, I would watch her run, a small thin figure in an impossibly large coat racing, racing to get to her job.

There was a perk that came with life in the Catalog Department: All the latest books crossed her desk, the books that drew the critics' attention and were the talk of literary New York. It was her job to help make sure they were sorted and processed to be divvied up among Brooklyn's vast network of sixty branches. She'd consult the massive tomes of *National Union Catalog,* published by the Library of Congress, to figure out the edition of a particular work. She mastered the Dewey decimal system and learned how to search through the four volumes of *Dewey Decimal Classification.* She was at ease using the many bilingual dictionaries on her desk, her facility in languages coming to her rescue. She'd also type up the little catalog cards that made a book an official part of the library system.

She was amazed at the range of works she was handling—not simply recent bestsellers but, for example, a 1950 edition of *The Complete Works of Homer, The Iliad,* or *The Secret of the Hittites: The Discovery of an Ancient Empire,* published in 1956, the year I was born, or *Alfred Hitchcock's Spellbinders in Suspense,* out since 1967.

She approached even the most menial task with relish. As she cataloged a volume of poems by Pushkin, for instance, she'd find herself reading it and taking notes. She might have been running the entire Brooklyn Public Library, she was so passionate about her work.

One evening Mom came home raving about a young author named Philip Roth who had published a splendid novel. She had been reading *Portnoy's Complaint* on her coffee breaks and found it "très risqué," she said with a chuckle, as she handed me a copy, "mais formidable." Another time, it was a new novelist, Joyce Carol Oates, who caught her eye. At home, she talked and talked about Joyce Carol Oates and wouldn't rest until I'd promised to read *Expensive People.*

Once I started it, I was horrified yet I couldn't put it down. It was a novel about a child who kills his mother and confesses to the crime, but no one in his affluent, well-heeled little town believes him.

At last, Mom felt connected to the world of literature she had coveted and that had eluded her most of her life. Not since Cairo and her work with the pasha's wife had she enjoyed the literary companionship she now found every single day at the Catalog Department of the Brooklyn Public Library.

*B*erkeley was ruled by a charming, elegant, and utterly dictatorial headmistress named Mary Sue Miller. Mrs. Miller was personable but rather terrifying. It was her first year and she had grand hopes for Berkeley, to liberalize it and bring it to a new era. But that didn't stop her from looking after the teensiest details or enforcing its ancient rules. She cultivated us, got to know us one by one. In that sense she was very different from the principals at my elementary school and junior high—cold, distant figures who sat closeted in their offices, emerging only for assembly or graduation. I had the sense that Mrs. Miller knew my particular strengths and my flaws. I didn't think she liked me very much.

When Wendy appeared in school in culottes, Mrs. Miller promptly sent her home for violating the dress code. She was constantly on patrol, and I noticed her walking down the hallways and darting in and out of classrooms. She wielded absolute control over every aspect of the school and of our lives.

Or so it seemed at first.

By my second semester, events outside were being felt within our little classroom. I wasn't sure anyone—even the formidable Mrs. Miller—was really in control. Fashion was changing, music was changing, the culture and even the mood of America was changing; and with the mounting anger over the Vietnam War, there was nothing sedate about what was taking place.

Betsy Raze, one of the brainiest girls in the class—and the most outspoken—arrived one day in a see-through blouse and no bra.

Technically, I suppose, she hadn't violated the dress code, which didn't have a clause about bras or gauzy blouses. To our amazement and hers, she got away with it. She wore the blouse again and again, sometimes with a short frosted brunette wig she fancied, and always with high heels, so that she appeared considerably older than her fourteen years.

The feminist movement was reaching a critical point. Traditional ideas about marriage and children were under siege and a new bestseller, *The Population Bomb*, was causing a stir at Berkeley, where even the head of the lower school raved about it. The book, by Paul Ehrlich, made dire predictions about what would happen if people had too many children and urged drastic action, including sterilization, to restrict the size of families to no more than one or two kids. I remembered how excited I'd been over these ideas years back at the Shield of Young David, and Marlene's contemptuous reaction.

Marlene was now married to a handsome Israeli named Avi. She and her husband had settled in Brooklyn, not far from the Syrian community of our childhood, with hopes of starting a family. I felt a million miles away from her—part of a new world, a world that was drastically different from the one she and I had once inhabited behind the divider.

When Berkeley decided to hold another dance, the rules that had tied us up in knots months earlier seemed antiquated . . . irrelevant.

Since our first mixer, pants with flared legs had replaced skirts and dresses for formal occasions. The dance was held in the gym, a large unsightly room that wasn't at all elegant or gracious like the assembly hall we'd used for the fall mixer. When I walked in wearing my demure little pale blue dress—the one from my date with Henry Finkelstein—I felt awkward and uncomfortable.

The music had also changed—it was much louder, much harsher.

Psychedelic rock and heavy metal had replaced the more melodious sounds of earlier in the year, the Fifth Dimension and Jay and the Americans swept aside by Frank Zappa and Jimi Hendrix and Janice Joplin, and I found the new rhythms—the revolution that Woodstock had wrought and that had finally reached us at 181 Lincoln Place—unpleasant and discordant.

A few of my classmates were smoking. I felt disoriented by the music, the atmosphere, the couples that had suddenly formed all around me. I didn't see Mrs. Miller anywhere and for once, I missed her, missed the order she imposed wherever she went.

I left without mingling with a single boy from Poly Prep or any other private school. I wanted only to be back in my room with my Charles Aznavour records, or listening to Dalida with Mom.

And suddenly, a social movement that I'd supported ardently from afar, that had seemed appealing from the vantage point of the women's section in my small synagogue, began to seem frightening and disquieting. The more strident and militant women sounded, the more freedoms they embraced, the more I wanted to retreat. But retreat to where? Even this elite private girls' school was turning out to be as turbulent as every other corner of American society at the dawn of the 1970s.

It was as if the world no longer had any safe harbors or women's sections or dividers.

Literally so, as I learned the day I went back to visit the Shield of Young David. I wanted to see for myself what had happened since its sudden closure several months earlier—if anyone had taken it over. I managed to enter the deserted building through a side door that had remained unlocked. Downstairs, in my old Hebrew school, the classrooms were still there, untouched, eerily silent. I made my way to the basement and a room that had haunted my childhood. It was usually locked, but once or twice my friends and I had been able to sneak in, always amazed by the sight of what looked like a swimming pool. It was actually an old ritual bath—a *mikveh*—where married women were supposed to go once a month and immerse themselves in a symbolic act of renewal, a process that made them "pure" again.

The room scared us, my friends and me, and we'd stayed away from it as children, not really understanding its purpose. Now, with the door left ajar, I decided to peek in. The lights were out; the water had been drained; there was refuse at the bottom of the pool. The room looked otherworldly, as if there still lurked the ghosts of bathers who had gone over the years to be purified of the sin of being female.

Finally and with some trepidation, I made my way up the familiar staircase to the sanctuary. It looked completely forlorn. The chairs were gone; the Holy Ark was gone, the bookcases that lined the wall had been emptied of all their books—no more copies of *The Kiddush Cup That Cried*, or Rabbi Avigdor Miller's attack on Darwin's theory of evolution.

And the women's section—it, too, was gone.

The wooden partitions had all been torn down; stray pieces were scattered in the back. The sanctuary was open and airy, but, alas, no one was there to enjoy it. Once upon a time, I realized, the mere thought of those dreaded barriers coming down would have filled me with glee. But I felt strangely desolate seeing the divider reduced to dozens of pieces of wood, ready for the trash. I wished that I could gather them and hold on to them.

*W*hen Miss Benson announced that Berkeley would be staging *Ring 'Round the Moon*, a play by Jean Anouilh, it seemed natural I should audition: The drama teacher herself urged me to read for her. But I didn't dare get my hopes up—I expected that all the major roles would go to upper-class girls. To my surprise, when the cast was posted a few days later, there was my name at the top. I had been given the plum role of Isabelle, the lovely young ingenue.

I hadn't been so excited since my last acting stint as Haman, the Evil One, in my temple's Purim *spiel*.

This was the major production of the year—boys from Poly Prep were going to play the male parts.

Miss Benson made it clear she believed in me: I was going to be a star. But rehearsals would take place after school and could last well into the night. Our parents had to give their explicit permission for us to be in the show. When Mom heard about the late sessions—with boys, no less—alarms went off. Once again, she huddled with my brother; the two decided that no, I could not accept the part, it simply wouldn't be safe.

I couldn't seem to argue with either my mom or César about their decision. They weren't budging, and nothing I said had any impact.

I didn't have the courage to defy them.

My mother was enjoying her newfound authority. With César supporting and guiding her, she displayed a stern side I hadn't seen very often. I realized that I missed my father's involvement in my life. Despite his legendary toughness, I had always considered Leon to be fair, and he tended to be mild with me. But he was in retreat now, the Invisible Man. Nothing any of us were doing—me in my high school agonies, Isaac away in college, Suzette off to Miami, Edith working—seemed to matter much to him anymore.

Mom's refusal to let me star in the play took the school aback. In an extraordinary step, Mrs. Miller sent her a letter urging her to allow me to join the production. She even promised that I and the other girls would be escorted to the subway every evening by the Poly boys. It was to no avail. No one, not even the headmistress, seemed to have any influence on my mother. Another girl—a sophomore—was tapped for the lead.

I was so crushed I refused to see the play.

In the weeks that followed, my mood, my entire outlook, seemed to change—I felt restless and agitated. I wasn't sure I wanted to be at Berkeley anymore. While I still got along famously with teachers like Miss Benson, I clashed with others. I had testy relationships in French class and gym, where I found the teachers to be martinets. I bristled when I dealt with anyone who struck me as bossy—which, of course, didn't serve me in good stead with that ultimate boss, the headmistress.

I now skipped gym class whenever I could. We had moved on from fencing and archery and badminton to field hockey and exercises on uneven parallel bars and pommel horses that I hated. I'd pretend to be sick and spend the period in the infirmary, being comforted by the school nurse, who allowed me to lie down and go to sleep.

Even my English teacher, Mr. Gardine, noted how much I had changed. He told Mom that I "could do better work."

Despite my recent ennui, I had thrived at Berkeley. I was much more assertive than I had been in public school. Emboldened by the As I was receiving along with all the encouraging comments from faculty, I was outspoken and overly confident, with emphatic opinions on every subject on earth, including subjects about which I knew nothing. I secretly felt superior to everyone around me.

Mr. Gardine tried to rein me in, telling me again and again I was too opinionated and self-absorbed and intolerant of others.

It was to no avail.

I had always done well in small, sheltered environments, and Berkeley had brought me back to the way I'd been in the women's section—supremely sure of myself, an Egyptian princess in Brooklyn.

One day, my mom was summoned to the school for a meeting with Mrs. Miller.

"Your daughter is arrogant," the headmistress told her. Edith was taken aback—she didn't know what to say, but her instinct was to apologize immediately and profusely on my behalf. When she came home, she seemed terribly upset and called me into the kitchen. As she fixed herself some Turkish coffee, she began quizzing me on what I could possibly have said or done to make such a poor impression on *la directrice*.

"Loulou, des fois tu exagères"—Loulou, sometimes you really are over the top—she said with tremendous sadness.

In the final weeks of the year, as the weather grew milder, going to school felt as it had at the beginning—serene, otherworldly. Sometimes we would have classes on the grass in Prospect Park. I had once again this sense of being in an idyllic, shielded environment, protected from the rest of the world.

That sense of calm was shattered when news broke of the shootings at Kent State University. National Guardsmen had fired on students who were protesting the U.S. invasion of Cambodia, killing four and wounding several others and sparking nationwide protests. Most schools in New York closed for a day as a gesture of mourning. When Berkeley remained open, the students declared a strike and refused to attend classes.

Mrs. Miller was upset, but what could she do?

Senior girls sought and received permission to wear black armbands at graduation in solidarity with the fallen of Kent State.

As if mirroring the general mood, I felt angry and rebellious, too, and wanting to protest—but to protest what?

I had recently picked up an American expression: thanks for nothing. I would use it mostly on my friends, jokingly, of course, but with a slight edge. "Well, thanks for nothing," I'd exclaim, feeling very American as I walked away from them.

We had a French exam, and several of us were upset with the teacher. But of all the students, I was the one who felt a need to convey my displeasure. I scribbled "Thanks for nothing" at the bottom of the exam paper. I was very pleased with myself. *I put this teacher in her place,* I thought.

Edith was asked to come in, of course. The school was crisp and to the point: Berkeley was not renewing my scholarship.

My mother was devastated, but when she gave me the news, I felt curiously detached. It was a terrible setback, of course, but I refused to see it that way. I'd already decided I was ready to move on.

Fine, I'll attend New Utrecht, I told Mom coolly.

"Tu retournes a une école publique—quelle horreur," she said; A public school again—how awful. She looked more despairing than I had ever seen her.

It had taken Edith five years to get me into an elite private school only to see me lose it in the space of ten months. What was it about me and my siblings, she cried, that we always squandered opportunities? Was there a curse on our family?

She was almost talking to herself now, bemoaning how God had given her children "toutes les grandes qualités," all the assets needed to prosper and do well on this earth—intellect, beauty, charm, personality.

For all the good it had done any of us.

Look at Suzette, she said, with the rage that flared up whenever she discussed my older sister, frittering her life away in odd jobs from Queens to Miami, no family, no husband, no children, no home, building nothing, achieving nothing. And now me, losing Berkeley, forced to attend *cette sale New Utrecht,* this awful New Utrecht.

I listened, impassive, to her outburst. Then I tried with all the calm I could muster to say that her illusions about private schools were only that, illusions. After a year in Berkeley, I wasn't sure there were that many differences between American children, no matter their social and economic backgrounds.

The changes the late 1960s had wrought had seeped everywhere into the culture. Those idealized young girls of my mother's imagination—*les jeunes filles rangées*—the proper, demure girls of the navy blue blazers with gold buttons and pleated skirts who had haunted my childhood didn't exist. Once upon a time perhaps, in Cairo or Paris, I would have met these delicate, refined creatures, but they weren't to be found in New York City at the dawn of the 1970s.

I didn't address my mother's larger fear—that each of us carried a self-destructive gene deep within us, that we were beset by demons and destined to trip ourselves up, and that calamity always lurked around the corner.

This was a recurrent theme for Mom, especially when she was in one of her dark moods. I had grown up hearing about the Curse of Alexandra, my wondrous and tragic grandmother. In recent years, my mother spoke of my sister as if she, too, were cursed— plagued by bad luck.

It was troubling to find myself included in that pantheon of star-crossed family members. Truth was that I didn't really know why I was leaving Berkeley, and why—or how—I had self-destructed.

After the school year ended, Wendy Gold and I made plans to get together. On a sun-drenched day, we walked through the Brooklyn Botanical Gardens and explored the lush plants and flowers and talked and talked.

I felt so sad; it was dawning on me I probably wouldn't see Wendy again.

I arranged to transfer to New Utrecht High in the fall even as my mother carried on with her work at Grand Army Plaza. She was facing a major change of her own, one that she dreaded:

In 1970, the Brooklyn Public Library gave Edith her own cubbyhole with her nameplate. She had reinvented herself as a professional woman a quarter century after leaving her teaching post in Cairo.

The library announced that the Catalog Department was going to be moved to its own building many blocks away. The department would be lodged in a windowless, converted garage on Montgomery Street, a bleak area of warehouses and abandoned factories.

But once ensconced on Montgomery Street, Mom and her colleagues discovered some redeeming features to the move. There was more space; they weren't all cramped together the way they'd been at Grand Army Plaza. And come lunch, they could wander over to the Botanic Gardens, which were literally around the corner, and dine outdoors while enjoying the wonderful greenery. Occasionally, Mom would buy a small plant for a dollar or two and bring it home because even a potted lily on the windowsill *ça nous aidera a reconstruire le foyer*—would help in rebuilding the hearth.

One day, my mother learned that she was being given her own office. It was very small, a cubbyhole really, but it came with a nameplate that said "Mrs. E. Lagnado." It may as well have been a corner office with a view—she was so thrilled with her new status.

Ed Kozdrajski, the young cataloger Mom found so charming and unassuming, came to work with a sleek camera he had purchased; photography was his latest hobby and he asked Edith if he could take some shots of her. She was delighted to oblige. In one photograph, she sits regally at her desk, her head held high, her back ramrod straight. In another, she stands beaming in front of her nameplate.

Ed's images captured what my mother herself couldn't put into words—that once again her world seemed filled with possibilities. She felt as if she had been given a new life; she experienced that same sense of buoyancy and soaring optimism as that day when Madame Cattaui Pasha had handed her the key to the pasha's library.

BOOK THREE

Cities

of

Refuge

MONTREAL, POUGHKEEPSIE,

AND THE UPPER WEST SIDE:

1973–1975

14

The Shrine
on the Mountain

*E*dith had never flown before, yet here we were at La-Guardia getting ready to board a flight to Montreal. Mom was very anxious, but I knew somehow that it had nothing to do with the prospect of getting on a plane for the first time.

It was July 1973, and I had graduated from high school only a few weeks earlier, finishing with top honors and offers to attend a string of prestigious colleges, from Vassar to Cornell, Mount Holyoke and Barnard. New Utrecht had turned out to be surprisingly supportive, and I'd found teachers I loved and who nurtured me. Craig Jones, a novelist who taught English, offered me one-on-one study sessions on Flannery O'Connor. Jeanette Stern, who insisted on being called "Mrs. Stern" though the age of *Ms.* was upon us, was so passionate about Tennessee Williams she had us reenact scenes from *A Streetcar Named Desire* in our Bensonhurst classroom, and made me passionate about him, too. Yet true to the family dictum about the *mauvais oeil* shadowing us at every turn, a period that should have been joyful, when I was set to reap the rewards of years of hard work, turned into a nightmare.

It all happened literally overnight.

Back in February, on the night my childhood friend Celia was getting married, Mom suddenly realized that I wasn't well. As I put on my evening gown, Edith insisted I see a doctor to explain the strange symptoms that I had tried to ignore. I was suffering from crushing fatigue and had lost nearly twenty pounds; merely getting up in the morning had become an ordeal. For months I had paid no attention, caught up in the crush of being a senior, applying to colleges, and struggling to do well to merit a scholarship, which I needed, of course, as my parents didn't have any more means than when we'd arrived in America.

That same week Mom skipped work and took me to Maimonides, the local community hospital where we always went in an emergency, and where the shattering diagnosis was finally made. After admitting me, the doctors concluded I had contracted Hodgkin's disease, a cancer of the lymph nodes that tended to strike teenagers. My cancer was advanced, they said, and needed to be treated aggressively.

The news of my illness left us reeling. Neither Mom nor I could quite understand what had happened, or why it had happened. "On t'a donné le mauvais oeil," she'd say; someone gave you the evil eye. At the hospital one night, she stayed by my bedside stroking my hair and speaking to me in Arabic, only Arabic, which was very unusual for her. She had reverted to the language of her youth, the language of her Cairo alleyway. "Loulou, ya helwa," she kept saying; Loulou, my pretty one. But her words made me even sadder. I felt a thousand miles away from pretty; I felt a thousand miles away from *helwa*.

My brother Isaac heard that Memorial Sloan-Kettering in Manhattan specialized in treating Hodgkin's. My family decided it made sense for me to go there though it was a long commute, and the cancer center itself lacked the homey feeling of Maimonides so that I felt lost and scared there.

The last several months had been spent undergoing tests and more tests, until the Memorial doctors finally settled on a radical course—ordering me to submit to months and months of radiation therapy, every day except weekends. I had a team of doctors led by an oncologist named Burton Lee. Dr. Lee was a portrait of WASP imperiousness and utterly intimidating—and I adored him.

One day in his office, Mom had pleaded with him to take over my care. "Madam, my fees would drive you mad," he'd said, but then, watching her slump over, he'd added soothingly, as if reconsidering, "Madam, I will save your child." Ever since then my mother had walked around the house in a daze, repeating Dr. Lee's words like a mantra: "Madam, I will save your child," even as I'd wondered, is that what it had come to? Did I really need saving?

Now, I was getting a break for one precious week. The doctors had practically ordered us to leave New York, telling Mom that a getaway was essential if I was to cope with the rest of the grueling therapy regimen. They wanted me to rest while they assessed how well the radiation was working—if it was even working.

Edith and I were mostly hoping to forget the last few months. We were both still in shock. I was thinner than ever, and Mom seemed so exhausted. She'd run off to the library in the morning, then come home in the afternoon only to race out with me to my radiation sessions at Memorial. After more prodding from my siblings and the doctors, she agreed with the need for a vacation.

But a vacation where?

We hadn't gone on holiday since Egypt. We were never able

to afford a family getaway after Cairo, and, besides, what was the point? My siblings were grown and dispersed and on their own. Occasionally, Dad would splurge and the three of us would go on a Circle Line tour along the Hudson River. They were oddly idyllic, these cruises—I'd make believe we were on a luxury ship that docked in distant shores, though the farthest we ever got was West Point. I found the glimpses of the cadets in uniform thrilling.

Paris was a logical choice but completely beyond our means. We had no experience jetting off to Bermuda or the Caribbean as other patients we met at Memorial were doing. Atlantic City was too close and had become shabby and derelict.

Montreal was my brother César's idea. He had gone there one summer and found it congenial and delightfully French. Suzette offered to buy us two round-trip plane tickets, and suddenly we were making plans and Mom was asking the library for a week off, though I could tell she had a heavy heart.

I was traveling in style. I wore a red sweater with a plunging neckline my sister had bought me when she came to visit, and a blue miniskirt with accordion pleats and beige leather sandals with high wedge heels. Suzette now lived in California—Miami had proved to be as disappointing as Queens and Brooklyn—so coming to see us was harder than ever.

For months, her input on my illness had consisted of emotional phone calls to Mom and letters to various members of the family urging them to get second, third, fourth, and fifth opinions, pleading with my mother to take me to Stanford or the Mayo Clinic. Now that I was in the throes of treatment, she was determined at minimum that we take this holiday and insisted on buying us the tickets and accompanying us to the airport.

In a photo taken at LaGuardia prior to boarding, I sit with my legs crossed, smiling. I look confident, jaunty, even oddly healthy.

I was none of the above. I had become a master of disguise.

Loulou and Edith at LaGuardia airport, summer 1973, about to take off for Montreal on a doctor-mandated vacation.

Some of my hair had fallen out but I hid this artfully, as I did the tattoos and red and blue ink diagrams that doctors had etched on my chest and stomach and back as they charted the course of my treatment, pinpointing with lines and arrows where the rays should be directed. I couldn't bear to look at myself and took pains to camouflage as much as I could. I wanted to wear what other women my age had on, but I was terrified that someone would spot the strange zigzags and lines and dots along my chest and my back and wonder what on earth was wrong with me.

We were supposed to be away for a full week yet we hadn't made any arrangements, not even a hotel reservation. And we didn't know a soul in Canada, although Mom did have the phone numbers of some old friends from Egypt, the Haboucha brothers, Victor and George and Haim. We had last seen them more than a decade earlier in Cairo, when they still ran the little synagogue, "le Kottab" (the Schoolhouse), my dad favored around the corner from our house. They had led the prayers there every Saturday morning, formidable and resplendent in their white sharkskin suits. In the strange existential lottery that determined where we would go after Egypt, and that dispatched some of us to Brooklyn and others to Adelaide and others still to Rio and Johannesburg and London, the game of chance that ripped families apart by forcing them to resettle on opposite ends of the earth, the Haboucha brothers and their wives and children had all ended up near one another in Montreal.

Victor and his wife had assured Mom by mail that hotels were plentiful.

"Saint Catherine," the Habouchas advised. "You must find Saint Catherine."

Outside the airport at the taxi stand, we told the driver, "Sainte Catherine, s'il vous plaît."

He nodded as if he needed to know nothing more to take us to the holy woman—the patron saint of Alexandria—herself. We arrived at a bustling street with a vast array of restaurants and boutiques. On a quiet corner was a careworn Victorian-style bed-and-breakfast that our driver recommended as he helped us with our luggage. The innkeeper offered us a suite at a rate we were surprised we could afford. It was a dingy old *pension,* but our rooms seemed palatial with their tall ceilings, large wooden armoire, and even a four-poster bed I immediately claimed as my own.

Edith suddenly seemed more energized—more confident—than I had ever seen her in New York. As we unpacked I saw the blue-checkered three-piece suit that Suzette had bought her for the trip. It came with a jacket, a skirt, a belt, and a pair of matching flared trousers with cuffs.

My mother had never worn pants—not once in her life—but

as we changed to go out, I begged her to try them. She slipped on the pants good-naturedly, belted them, and—voilà. There she was transformed from a shy immigrant woman into a stylish, Americanized modern woman.

We wandered down Saint Catherine Street in a strange state of exuberance we hadn't felt in months. It was as if from the moment we had landed we'd found a vestige of our old optimism—a sense of hope again. The streets were vast and orderly and impossibly clean. I ran in and out of every stylish clothing shop along Saint Catherine with a newfound energy and vigor. Together, Edith and I peeked into the cafés and patisseries that were everywhere around us. Everyone spoke French and that in itself was comforting, as if we were back in Paris or Cairo again. But it was a French I had trouble grasping. My mother, on the other hand, seemed perfectly at ease with the dialect. She'd stop to chat with perfect strangers and ask them for advice or directions: They seemed delighted to oblige.

We had an unspoken agreement—not to bring up my illness. We were pretending to be carefree tourists on holiday, taking in some sights, having fun, and seeking a respite from the scorching heat of New York. And that worked to a degree. Montreal was such a delightful city, it was easy to lose yourself; or perhaps we were simply relieved to be so far away from Memorial Sloan-Kettering that we could put it out of our minds.

But then I'd remember some of the doctors' dark warnings and feel melancholy all over again. While there had been significant breakthroughs with Hodgkin's—even talk of a cure—my physicians warned of the consequences of the treatment. There will be sequelae, they said again and again, referring to the various ailments and complications that could befall me either as a result of my weakened immune system or the radiation itself. I could count on relatively minor setbacks, like getting more cavities, as well as being at risk for developing another cancer—breast cancer or even leukemia.

One of the worst "sequelaes" had already come to pass. Once I started treatment, almost immediately I lost the ability to have children, a fact that was so secretly shattering to me I told none of my friends and couldn't even talk about it while alone with Mom on this trip. Even if I miraculously overcame Hodgkin's, my future seemed filled with dark and terrifying possibilities.

But perhaps not in Montreal, and not this week.

At the hotel that first night, Mom sat down to dash off a letter to Suzette to say we had arrived safely and were getting settled. Montreal, she scribbled enthusiastically, "a le charme de Paris et la modernité de New York." The city had the charm of Paris and the efficiency of New York.

Since her childhood with Alexandra, Mom had always been superstitious about running into priests. "Un prêtre ça porte malheur," she would tell me—priests bring bad luck—then she'd take me by the hand and we'd cross the street to avoid them, exactly as my grandmother had done with her back in Sakakini. But we were now in a city of priests, a city of four hundred churches, of crucifixes and shrines and votive candles and modern-day pilgrims who prayed with abandon. We hadn't come to find God in Montreal, yet God was finding us, God was everywhere around us, and somehow that reassured us, made us more hopeful.

I was hungry again: that was the first sign. There were *creperies* on Saint Catherine that sold wafer-thin pancakes sprinkled with sugar or jam. I'd devour several of them at once while Edith watched delighted.

They tasted delicious, and nothing had tasted delicious of late. Shortly after we arrived, Mom telephoned the Habouchas and we were invited to come over. We made our way to a modern building in Côte-St. Luke, one of Montreal's tonier neighborhoods. In his apartment perched high up in the sky, my father's friend, one of his closest companions from Cairo, greeted us like a lord of the manor. Victor Haboucha had traded his white sharkskin suit for an elegant double-breasted navy blue blazer with gold buttons, tailored slacks, and sleek, expensive shoes. He looked jaunty and debonair—as if being uprooted and transported thousands of miles to a land of cold and snow hadn't fazed him one bit. He and his wife gave us a tour of their home, which struck us as the height of luxury, a far cry from our modest ground-floor digs in Brooklyn.

Mrs. Haboucha motioned us to a door that led to a sunroom filled with plants and flowers.

This was "le terrarium," she declared. Mom looked momentarily stunned.

"Le terrarium?" she repeated, and Mrs. Haboucha nodded, her superior social status established beyond the shadow of a doubt.

Victor was my father's age; his wife wasn't much younger than Mom. Both had been forced to reinvent themselves in a culture that was alien and foreign. Yet they seemed to have found their way. Victor, for example, was still a boulevardier; he simply frequented different boulevards than the ones he'd loved in Cairo.

dith, I discovered, was a methodical traveler. She wasn't content as I was simply to stroll up and down Saint Catherine poking my head into the fashionable boutiques; Mom wanted to consult guide books, to explore, to take in as many of the sights as we could handle. Commanding in her pantsuit, she would lead me by the hand to the city's center where the tour buses gathered.

We took guided tours constantly, obsessively. There were two-hour tours and four-hour tours, tours with multiple stops at shopping venues and tours without any stops at all, where a large bus simply whisked us from one tourist destination to another, and a guide kept up a loud stream of chatter as Montreal dissolved into a blur of churches and more churches. It was a city that believed in miracles.

And we were in the market for a miracle.

Each one of our tour buses began with a visit to the Old City, and then, after a requisite stop at Expo 67, the site of the old world's fair, they'd begin the climb up and around Mount Royal for a closer look at the luminous steel cross that dominated the city, the cross that we could never seem to escape.

The tours all ended on a high slope of Mount Royal, the grounds of a sprawling modern-day shrine called Saint Joseph's. We were in the Lourdes of Montreal, our guides informed us, where thousands of desperate travelers—the sick, the infirm, the disabled—journeyed each year. They would climb on their hands and knees up the hard stone steps to the shrine in hopes of finding a cure. No disease was so virulent, no condition so crippling as to be beyond the healing powers of the Shrine on the Mountain.

I noticed that no matter when we arrived—early in the morning, late in the afternoon—the church was thronged. Inside a small chapel, thousands of red votive candles were burning, and there was an overpowering smell of incense, incense so thick it was difficult to breathe. In a small crypt behind the chapel, Brother André, the oratory's founder, lay buried; a glass case nearby contained his heart.

There was also tangible proof that miracles had taken place. Tucked away in discreet alcoves were hundreds of ancient canes, decaying crutches, tarnished steel braces, walking sticks. They had been left behind years, even decades, earlier by invalids who had made the journey to the shrine on the mountain. In some corners there were so many of them they reached all the way to the ceiling. I leaned over to touch a cane, as if that would dispel any doubt I still harbored about the possibility of a godly intervention.

We had spent months hoping for a miracle, Mom and I, only to hear there were no miracles to be found. We'd had no choice but to put our faith entirely in science and medicine. Doctors, hospitals, tests, procedures, oncologists, radiation machines, radiotherapy sessions—those were our salvation, we were told. There were no shrines in New York and no one left who believed in them anyway.

Standing there, looking at the altar with its ten thousand red candles, we were believers once more, every bit as much as the pilgrims who wept at Brother André's grave, or the visitors who kissed the glass case that held his heart, or the Christian soldiers who crawled up the one hundred steps leading to the oratory on their hands and knees in supplication.

At night the city changed. God went home and his churches were locked shut; the memorial candles flickered out and were replaced by the lights of a thousand restaurants and cafés and nightclubs. Saint Catherine, the saint of martyrs, now presided over a crowded streetscape of tourists and summer revelers.

Our last night in Montreal, we didn't want to go to sleep. We wandered restlessly from one café to another, one creperie to the next. Neither of us could bear the idea of going home to New York City.

Neither of us wanted to consider what awaited us once we got there.

A theater on Saint Catherine Street beckoned; ticket buyers were queuing up to see the new 007 film, *Live and Let Die*.

I was hesitant at first. Back in New York, my family had tried taking me to the movies, but I couldn't sit still. Instead of the images on the screen, my mind would drift to my treatment, and

I wanted only to flee the theater. Besides, neither Mom nor I were great James Bond fans. But the theater was crowded, it was a lovely, balmy night, and we were still in the same state of exhilaration that had overtaken us from the time our Air Canada flight had left New York, even now that we knew we were heading back.

At some point halfway through the film Edith leaned over in the dark. "This should be your motto," she whispered.

I looked at her not quite understanding.

"Live and let die," she repeated, and she was smiling, yet I could tell she was serious.

"Loulou, that has to be your philosophy from now on."

And that was how my mom addressed the darkness of the prior months, by urging me to heed James Bond.

She didn't elaborate, not then and not later, when we returned to our hotel room. She was, I realized, telling me that I had to change—that if I had any hope of surviving, I couldn't grow up to be like her, timid and meek and passive.

My mother had led a life of sacrifice. She had sacrificed herself for my father, had abandoned her dreams to marry him, had given up the key the pasha's wife had handed her and all the doors it would have opened. She had gone on to devote herself to us, her children. But as far as she was concerned, my illness was enough of a sacrifice and she was telling me not to be like her, not to give up my hopes and ambitions. I had to become tough and even ruthless—I had to live and let die.

She packed silently, neatly folding her pantsuit and putting it away at the bottom of the suitcase.

We woke up the next morning and caught our scheduled flight back to New York.

· 15 ·

The Fall of Afterward

At last, the summer of 1973 was over and I was supposed to start college. Back in the spring, when I had settled on Vassar, I wasn't convinced I would actually be able to attend—I felt too weak and too sick, and my treatment seemed endless. I was sure the doctors were lying to me. I had a feeling I would be there at Memorial for years, submitting to the unrelenting daily regimen of blood tests and radiation, blood tests and then more radiation.

Suddenly late in August, I was told I was free to go: I had completed my therapy, the doctors declared. I weighed ninety-eight pounds—twenty pounds less than I had a year earlier, and I was crushingly tired. I could barely walk a block or climb a flight of stairs without feeling that I would collapse. How was I going to manage on my own at school, without Edith helping me?

Besides, I wasn't ready.

I hadn't the foggiest idea how to dress or prepare for Vassar. My fashion consultant was my brother Isaac, who felt the only store that sold appropriate clothes was B. Altman's. Off we went together to the cavernous building on Thirty-Fourth Street. I was a stranger to the great Fifth Avenue department stores—when Mom and I shopped, we favored discount outlets.

As I walked with Isaac up and down Altman's elegant and largely deserted floors, I realized that I could afford nothing—not the wool crew neck sweaters he seemed to think were essential to my wardrobe, or the cotton turtlenecks that went with them, or the corduroy jeans, or the one accessory he pointed out in a hushed tone as the most important of all: Pappagallo shoes.

My brother fancied himself a connoisseur of high prep.

As far as he was concerned, I had to have a pair of Pappagallos.

They were located in a special corner of Altman's called the Shop for Pappagallos. This small fanciful universe had the look and feel of an exclusive private boutique. There, all along the floor and on small shelves, were little felt handbags with wooden handles, belts that had gold buckles in the shape of seashells, silk scarves, as well as the famed signature Pappagallo loafers. Made of soft Italian kidskin, they came in every color scheme imaginable—pink and mustard yellow and lime green and navy blue and crimson red.

I stared at them at first slightly disoriented, not quite understanding all the fuss. They looked nothing like the shoes my friends and I were used to wearing in Brooklyn—chunky, clunky, aggressive high-heeled numbers that were either brown or black. We didn't wear flats at New Utrecht High, and we certainly, certainly didn't wear pastel pink flats.

Besides, I didn't have enough money to buy the simplest pair. My brother seemed impatient. If I was really going to attend Vassar, I needed to wear Pappagallos.

I pictured an armed guard standing at Main Gate, peering at my feet to see if I was wearing my usual budget shoes from Miles on Eighteenth Avenue, and frowning; I didn't really understand high prep and its meticulous, ironclad rules—not yet—but I knew instinctively that I should heed my brother.

I finally purchased one navy blue wool crew neck Isaac assured me would at least pass muster. And that, along with careworn jeans now a couple of sizes too big and some blouses bought on our trip to Montreal, made up my college wardrobe. I packed to leave for Poughkeepsie.

When I went back to Memorial for one final checkup, I ran into Elena Barnet, my social worker. She and I were very close: She had sat with me day after day as I awaited my treatments. I decided to confide in her my angst over the Pappagallos. She seemed both amused and sympathetic and urged me to buy at least one pair. I nodded but I knew I couldn't swing it. I would have to start Vassar without Pappagallos.

Then it was time to go. We went to Grand Central and boarded the train to Poughkeepsie, Mom and Isaac and I. Each of us was holding a piece of luggage or a parcel. We took a taxi to the campus and made our way to my dormitory.

I had been assigned a room in Noyes, a low-lying modern structure built in the late 1950s that was nothing like the gracious nineteenth- and early-twentieth-century Tudor-style buildings on the rest of the campus.

It certainly wasn't the living arrangement I had envisioned. Isaac had taken me to visit Vassar one day in the summer, a day when I'd been granted a rare reprieve from my radiation treatment. We had passed through stately Main Gate and walked onto a lush, verdant green campus with massive fir trees, towering pines, and flowering bushes. A few stray students were lounging on the grass, reading. The library looked almost like a cathedral with its vaulted ceilings and massive stained-glass window.

I thought that it was the most beautiful place on earth.

I had actually fallen in love with Vassar before I ever stepped foot in it, a captive of its mystique. This was, after all, Jackie Kennedy's Vassar I would be attending, Jane Fonda's Vassar, the Vassar of Edna St. Vincent Millay, the Vassar that had shaped and nurtured Mary McCarthy. On my first visit, I'd wandered into the library and asked for the yearbook of the Class of '33, McCarthy's class, and sat there staring and staring at the photo of the pretty young woman with the slight ironic smile. I couldn't wait to sip sherry with my professors and nibble on Vassar Devils, the chocolate dessert smothered in vanilla ice cream, marshmallows, and fudge sauce that was served exclusively at Alumnae House. I was all set to dress in pink and gray—the school colors that symbolized the "rosy dawn of women's education piercing through the gray clouds of ignorance." These were all pleasures that awaited me, according to the mounds of college literature postmarked Poughkeepsie that kept descending on Sixty-Fifth Street.

Reading McCarthy's *The Group* in high school had given me a sense of a charmed, almost magical environment where elegant young women shared rooms in the Main Tower, then came together every afternoon to mingle over sherry and biscuits in the beautifully apportioned Rose Parlor.

Noyes—modern, functional, deeply antiseptic—didn't have a Rose Parlor and didn't, to my knowledge, serve sherry. I had been assigned a single room upstairs, in the back of the second floor; my window, I noticed to my shock and dismay, overlooked a small cemetery.

Unlike my classmates who were moving in with enormous trunks and open crates overflowing with rugs, bedspreads, stereos, curtains, even small refrigerators and TV sets, and whose parents were helping them unload their belongings, I had brought nothing except what we could carry—my clothes, two sheets, a blanket, and a pillow. I had a portable manual typewriter, an

Olivetti, that was my prized possession, and I thought it very wonderful until I noticed that everyone around me was lugging massive electric and Selectric typewriters.

I looked at my little blue-green Olivetti; suddenly it didn't seem quite so formidable.

Then I sunk down on the bed, not much larger than a cot. My room struck me as impossibly bleak and bare.

"Tu est sûr que tu veux rester?" Edith asked, sensing my dismay. Are you sure you want to stay?

Did I have a choice? I shrugged. She and my brother left and I was alone.

I wandered downstairs to the lobby. I was suddenly starving, but I had no idea where to eat. I asked the lone person I spotted outside, a boy, a freshman like me, resplendent in white tennis shorts and polo shirt, with black sunglasses. He amiably pointed me in the direction of the central dining hall.

Vassar had a long tradition of dining rooms in every dorm, each with its own special culture and personality. The dining rooms were part of the school's charm and had figured prominently in McCarthy's novel. I imagined them as cozy and intimate, where I'd wander down to breakfast in pajamas and a bathrobe. I had no idea they had been abolished in a cost-cutting move the year before to make way for the stark efficiencies of "ACDC"—the All-College Dining Center.

I arrived at a large, imposing building with white columns. Inside, it was sheer chaos—crowds of students with trays in their hands, queuing up on long lines that fanned out in different directions.

"Where is the kosher section?" I asked a woman who appeared to be in charge.

I had assumed there'd be a special place for me to eat, stocked with the kind of food I liked.

Loulou as a freshman at Vassar; dining card for the college's central dining hall, fall 1973.

The woman seemed taken aback. Perhaps no one had ever asked her this question.

"We don't have kosher food," she replied, eyeing me curiously as if I were a visitor from another planet. Suddenly I felt like one. I was completely stunned. I had been adhering to a kosher diet as far back as I could remember. When I'd picked Vassar, it never occurred to me I wouldn't be able to eat my normal fare. I was used to having kosher food around me everywhere I went.

I hadn't figured out yet that the rest of the world wasn't like my little corner of Sixty-Fifth Street.

"There is cottage cheese and salads over there," she said helpfully, pointing to an area off to the side.

My first night at Vassar I dined alone on a bowl of cottage cheese and wondered what I had done and how I would survive.

As I sat surveying the scene, I realized that many of the freshmen seemed to know each other—no doubt graduates of the same upscale suburban high schools and boarding schools.

Some of the Vassar women were very striking, with shoulder-length sandy hair they wore swept back with a headband. They sat at large tables, seven or eight of them, speaking intently with one another. Most wore polo shirts or white turtlenecks tucked under a sweater.

There they were, the living embodiments of the Group. But I didn't feel especially happy to have found them.

Breakfast, at least, felt hopeful.
Unlike lunch or dinner, there were ample foods I could eat—stations where they made eggs or, occasionally, pancakes, shelves stacked with small boxes of every cereal imaginable, fruit salads, blueberry muffins, and the ubiquitous bowls of cottage cheese, each decorated with a maraschino cherry in the center. There was an abundance to these morning displays that I relished. I figured that if I ate enough at this one meal, at least I wouldn't starve to death while pursuing my Vassar degree.

I developed a passion for scrambled eggs, which I hadn't tasted before—Edith didn't make them at home. To me, they were a piece of Americana, a portion of that exotic culture I craved and that had propelled me to choose Vassar in the first place.

I'd pile several ladlefuls of scrambled eggs on my plate and look for a table.

The central dining hall was typically deserted in the morning. Few students came out for breakfast, and those who did tended to sit by themselves, quietly reading the *New York Times* as they sipped coffee from their large Vassar mugs.

Breakfast wasn't a social occasion, not the way the other

meals were, and it was a relief to sit alone without feeling self-conscious.

Although the year was only starting, I looked around me and saw that groups had already formed. Everyone had their circle of friends with whom they dined day in and day out. But how they'd come together was a mystery to me. I could never penetrate the magically protective bubble these groups created around their chosen members.

It didn't help my social status when ACDC's manager approached me one day to say she'd found a solution to my kosher conundrum. Vassar, she proudly informed me, was prepared to order frozen kosher TV dinners. The meals would be from the same company that catered to the airlines; the central kitchen would heat them for me every night. The manager seemed solicitous, and I realized that she was trying hard to be sensitive.

Dinner, already an ordeal, took on a nightmarish cast.

Every night, I would report to the back of the kitchen and be handed a large silver package, steaming hot, that I lugged on my tray. As I walked through the dining room holding my little kosher TV dinner, looking for a table where I could sit, for people I could join, I noticed other students staring at my tray, puzzled.

The food itself was barely edible—some form of meat or chicken with a little portion of noodle pudding or a side of overcooked vegetable.

I wondered if I'd be better off simply eating cottage cheese.

A few weeks after school started, I realized that I wasn't well. I couldn't stop coughing.

I felt a suspicious lump near my left shoulder.

I was sick again, I was sure of it. My Hodgkin's had come back, I thought, panic-stricken. I raced over to Baldwin, the infirmary, where a cheery woman physician immediately saw me. I told her my history, the fact that I'd been treated for cancer shortly before coming to college.

"Where is the lump?," she asked as she examined me; she seemed nervous, too. I pointed to a knob by my shoulder. She chuckled. It was simply a bone, she assured me. I'd lost so much weight that virtually all the padding around my arms and chest had vanished.

I coughed some more. "Do I have pneumonia?," I asked her anxiously. I feared the worst. In my months at Memorial, I had noticed that patients didn't necessarily die of their cancer—they tended to become so thin and weak and emaciated that they'd contract pneumonia and never recover.

"You are fine, but take it easy," she replied thoughtfully.

Some time later, I was back at Baldwin, feverish and unable to stop coughing; I was extremely tired.

The doctor decided it was prudent for me to stay overnight. I was taken upstairs to a room with a small cot. It was exceedingly quiet, especially in the evening, when most of the nursing staff would disappear. From my window I could see the campus with students walking in pairs or small groups, chatting. I found myself wondering once again if I would survive Vassar.

The kosher meals followed me even to the infirmary. Somehow, the nurses had been instructed about my diet so that now I was being offered the reheated TV dinners twice, for lunch as well as dinner.

The first couple of days, I slept constantly; all I wanted to do was sleep. One afternoon, I dreamed that I was back on Sixty-Fifth Street, with Edith taking care of me. I woke up to find her there standing at my bedside peering at me: Had I been hallucinating? How did my mom, who at times struck me as so helpless, who had seemed more confused and befuddled than ever since my illness, manage to find her way to my infirmary room on the edge of a campus where I myself still felt lost? Simple, she explained: When Vassar called to tell her I was ill, she had taken the train by herself from New York to Poughkeepsie and then a taxi to the gate, where she had asked enough students until she'd found the building.

Edith stayed by my side that afternoon, stroking my hair, and then left as I slept to catch a late evening train back to Brooklyn.

I lived in my blue B. Altman's sweater. The sweater was one or two sizes too large and I found it oddly comforting. I would slip it over my jeans, sometimes without even bothering to wear a shirt underneath. It was like wearing body armor—or maybe a potato sack.

The pursuit of beauty, one of my primary objectives for so many years, going back to my fascination with Emma Peel, was over. My arrogant years had officially ended in the summer of 1973.

There were no visible scars from the treatment, but that was almost beside the point.

I wasn't the same. As a little girl, I had been deeply vain, always looking at myself in the mirror, posing this way and that, pretending to be Mrs. Peel. Now, it was as if I couldn't even bear to look at my own reflection. As I grew up, I'd favored edgy clothes in lively colors, sexy dresses, low-cut sweaters, tight jeans. I wore my hair long and loved how it flowed so luxuriantly past my shoulder.

Though it had grown back even thicker than before once the radiation ended, to my eyes I looked nothing like I did before. I didn't feel in the least bit attractive. Actually, I felt . . . invisible. The true "sequela"—the aftereffect the doctors had failed to mention—was that I had been completely diminished by my illness. I felt inferior to every woman I saw at Vassar—it didn't matter if she was a plain Jane or an aristocratic beauty out of the pages of Mary McCarthy. The feeling, never articulated—perhaps not even to myself—was this: She has what I don't have. She can have children and I never will.

Life was now simply about fitting in. That was the essential— to blend, to mimic, and to resemble the other women on campus.

But no matter how hard I tried, I realized that I wasn't a bit like them. No matter that I, too, now wore crew necks and turtle-necks. I would observe the girls walking on the quad or reading in the library and realize that what I couldn't approximate was their easy elegance, how even at their most casual, it was clear they were children of privilege and I was not.

I tried to find comfort in my work. I had always been in my element in a classroom—never afraid of voicing my opinions. But suddenly, I felt fearful—so fearful that I stayed silent, even when I most wanted to speak up. I listened as my classmates spouted whatever came to their heads, but I could never so much as raise my hand or volunteer an answer or offer a point of view. When-ever I tried to speak, my voice seemed alien and almost disem-bodied.

I worried about sounding foolish, no longer sure of my ideas. The self-confidence and drive that had propelled me since I was a child were gone. My exuberance, my enthusiasm—my arrogance— had all seemingly vanished in the Hudson Valley mist.

Instead of focusing on what my professors were saying, I found myself staring at my classmates' watches. My eyes would wander from one arm to another. I was fascinated by how many wore gold watches—solid gold, gleaming, opulent. Even in a classroom, where everyone took pains to look casual in their jeans and sweat-ers, there seemed to be small but significant displays of wealth.

When I was in the hospital months earlier, my brother César had bought me a watch as a gift one night. It was a Tissot, extraor-dinarily sleek, silver, with a blue suede bracelet. I had loved it and taken it with me to Vassar. Yet here, my Tissot seemed to lose its value amid the softly gleaming golden watches of the other Vassar women.

When I didn't look at my classmates' wrists, I stared at their feet.

Where were the Pappagallos?

I spotted a few of them. Mostly though, along with the crew necks and jeans and corduroys, students favored not the slender loafers my brother had urged me to buy but thick, odd-looking maroon shoes with white rubber soles they called "Top-Siders." I'd never seen them before. I didn't know that they were boat shoes, emblematic of the lifestyle many of my classmates enjoyed beyond school—a life of summer homes and sailing expeditions that was completely foreign to me.

I finally scrounged up the nerve to ask a friendly girl in my dorm where I, too, could get a pair of Top-Siders.

"I order mine from a catalog," she replied helpfully and offered to lend it to me.

But as with the Pappagallos, I couldn't afford the mail-order Top-Siders either, and I continued wearing my discount shoes from Eighteenth Avenue, the shoes that I hoped would pass muster and, of course, didn't, because there was a tyrannical element to the rules of high prep and individuals' social status; their ability to make friends or not, and the kinds of friends they made, could be determined swiftly, ruthlessly, by glancing at their feet.

I preferred to travel incognito in this new world of mine. I favored dark clothes, dark shoes, and a long dark woolen coat. But I couldn't help wondering whether my new affinity for grays and blues and blacks—in contrast to the luminous reds and greens I had always loved—was in some way connected to the previous summer, the summer that had been devoid of any color, any light.

Side by side with the rigid rules, I was glimpsing a freer and more open lifestyle than I had ever known before. There was a strong, almost militantly feminist culture that was flourishing at Vassar, and it made me uneasy. The English Department, I learned, had multiple Sylvia Plath scholars. It was as if the only book worth reading, studying, and analyzing in the fall of 1973 was *The Bell Jar.*

In my dorm, some of the freshmen women had boyfriends who visited them on weekends and stayed with them—two levels of freedom I had never known and couldn't even have imagined prior to coming. I was secretly shocked, but couldn't really confide in anyone: My classmates all seemed to take it as part of the natural order.

There were also women who were openly gay, and they enjoyed a kind of lesbian chic. They had their own groups and they, too, enshrined Mary McCarthy. They were often conjuring up Lakey, the novel's beautiful lesbian character, as a role model.

They all terrified me—the women with girlfriends, the women with boyfriends. I kept my distance from both.

*I*t was very gradual, my decision to run away from Vassar. It wasn't really conscious, at least not at first.

At first, it was simply a matter of catching a slightly earlier train Friday to go home for the weekend. No point lingering in Poughkeepsie, a hopelessly run-down little city, I thought as the weekend drew near and my classes ended. The idea of staying put, of going back to my dorm and trying to unwind and make new friends, held little appeal for me. I dreaded my bare room, which still lacked curtains and a rug. I didn't like to look out my window—more and more the cemetery below filled me with terror. Being stranded in this room for an entire weekend was unbearable. It made sense to head home Friday the minute I was done, to rush from my last class to catch a taxi to the train station.

I felt more of an outsider with every passing month.

By the late fall and winter, I had made inroads with only a handful of students. I had befriended a group of girls who came from, of all places, San Francisco high society. A few had attended the same tony boarding school—the Santa Catalina School for Girls—and they lived near one another in Pacific Heights. They had wonderful names; my favorite was Lucia Blair, whose bearing matched her aristocratic moniker, yet who still managed to be remarkably sweet and friendly and down-to-earth. I had a strange feeling that, like me, they weren't in love with Vassar—that it had somehow disappointed—and they didn't feel a part of it either. They obeyed the rules of high prep, but they did so with a kind of California mellowness, as if they preferred not to take any of it too seriously. Most wonderful of all was the fact that they seemed willing to let me enter their charmed circle.

By winter, a few like Lucia were excitedly planning their debutante balls back home. I was regaled with stories of the gowns they would wear and the coming-out parties that were being planned for them by their families. This was much more the Vassar I had envisioned.

The coming-out parties my friends were holding were like my Aznavour records—stirring, soulful, and hopelessly antiquated.

On campus all around me, students were wearing "Impeach Nixon" buttons. The favorite social activity wasn't sipping sherry but going for a beer at Pizza Town or Frivolous Sal's, the two popular local bars. There were rock concerts held on the edge of campus or in neighboring towns, and I had no interest in them, either. The only music I enjoyed was Aznavour and his songs of a prior generation. I would have loved that sherry the admissions literature had promised, but if it was being served anywhere at

Vassar, no one told me. There was a drinking culture and a drug culture but no sherry culture.

As a little girl new to America, I had watched with such longing the annual debutante ball at the Waldorf Astoria on TV. I loved listening to the emcee's flowery descriptions of the debutantes' *blanc d'ivoire peau de soie* gowns. Mom and I would stare at the girls with their handsome escorts; even the ones who were overweight or plain beneath their finery seemed to have wonderful-looking companions.

I had asked Edith if I, too, would have a debutante ball and she had assured me with her usual optimism that I would indeed. It made perfect sense to my mom that I'd have a coming-out party, that I would be introduced, as she liked to put it, *à la haute société*— to high society.

She expected nothing less from life, though life had bitterly disappointed. And perhaps because it had disappointed so completely, her dreams for me—her last hope on this earth—became more and more outsized, more and more outlandish.

Perhaps that is why she hadn't objected too strenuously at first to Vassar. She was torn between her grand ambitions on my behalf—which meant casting me off into the greater world, letting go to allow me to fulfill my destiny—and her passionate desire to keep me as close to her as possible in that four-room apartment on Sixty-Fifth Street, though she realized that meant I may not have much of a destiny at all.

I was only a spectator as my California friends chatted about their upcoming balls, much as I had been as a child, as cut off from the real-life galas being planned in Pacific Heights as if I were watching the International Debutante Ball with Edith on our TV set in Bensonhurst, both of us hypnotized by the splendor of the ballroom, the swish of the gowns, the loveliness of the young women as they dipped and curtsied across the Waldorf's Grand Ballroom.

*C*ome the winter break, we dispersed—my friends to their coming-out parties on the West Coast and me home to Brooklyn. The apartment felt smaller, shabbier, quieter. I found my father exactly where I'd left him, praying. He barely looked up when I arrived, refusing to acknowledge either that I had left or that I had returned.

Mom was still in her little half kitchen, making herself some Turkish coffee. With the years she had become exactly like her own mother, Nona Alexandra, who survived by drinking cup after cup of *café turc*, a ritual that went on late into the night.

Was she even bothering to eat?

I had finally put on weight in my months away—rather surprising in view of my daily struggles around meals. Somehow, between the hefty plates of scrambled eggs in the morning and the kosher TV dinners at night, I was nearly back to what I'd weighed more than a year or so earlier, before the illness had manifested itself.

Edith, on the other hand, looked frightfully thin. Always a wisp of a woman, she seemed to be disappearing. I suspected that she wasn't eating much beyond her beloved *bâtons salés*, the sesame-covered Stella D'oro breadsticks that were among the few items of food I found in the house these days.

In the past, the two of us would always dine together in the evenings. She loved to cook for me, typically a large pot of *hamed*, the stewlike dish where she mixed in small meatballs along with potatoes, carrots, artichoke hearts, and celery, the lot simmered in a lemony, garlicky, minty broth. It was such a fragrant dish, especially when she added the green mint leaves into the pot, that its scent would fill the kitchen.

With me gone, had Edith simply decided to forsake eating? Had she stopped making *hamed*?

But on this winter break, we fell back into our old routines. She cooked for me—and herself—as I regaled her with stories about life at Vassar, trying to be light and entertaining so as not to betray how I really felt. I wondered if she'd guessed that going away had turned out to be an enormous disappointment. She would gingerly ask me if I planned to go back to Poughkeepsie in January, in the same way as on that first day of college in September she had questioned whether I wanted to stay.

We both strenuously avoided any discussion of the previous year. She used only vague, elliptical terms when discussing my bout with Hodgkin's—it was always about *ta maladie*—your illness—and she would never, ever say "cancer." And because she couldn't even say the word, we couldn't talk about how I was doing, couldn't have a heart-to-heart on whether I was coping now that the crisis had supposedly passed.

Even more taboo was any mention of marriage. It had never been a popular topic in my house: My parents had failed so miserably with Suzette, unable to persuade her to settle down and have a family. As I grew up, my mom liked to pretend I was being groomed for a special destiny that would transcend what other women expected out of life. I looked at my friends—who even at a young age fantasized out loud about the weddings they'd have—with disdain. I was sure I was meant for so much more.

And my mother had fed my sense of superiority. She encouraged my arrogance.

Of course, now the dynamic had completely shifted. It wasn't a question of my choosing to get married; it was about whether I would ever be chosen. What Edith couldn't bear to say—nor for that matter could my father—was that I had become unmarriageable, at least under the unspoken rules of my Levantine community.

Childless women are unmarriageable in Middle Eastern culture, and that was the most taboo subject of all. Mom, who had raised me to spurn a more traditional life, wasn't going to dwell on the fact that it was denied to me now. My father, never one to tackle difficult subjects, avoided it altogether.

I would probably have silenced them anyway. Ultimately, both were, in their own awkward, misguided way, trying to be kind— they knew I was injured, perhaps even more than I realized, and feared treading on any ground I would find hurtful that would wound me even more than I was already wounded.

Even during the months of treatment, I'd had an inkling of the change in my social status. At some point, rumors of my cancer diagnosis had seeped out, and a friend of mine from the days of the Shield of Young David, Miriam, told her parents.

Miriam had been an odd bird at the Shield of Young David—a relatively recent arrival to our congregation, she was an Ashkenazi Jew, one of the few in our midst whose family traced their roots to Eastern Europe as opposed to the Levant. Overweight and awkward, Miriam had been a social outcast at a string of religious schools and Jewish camps. Our shul and women's section offered her a welcome embrace: That is what made the Shield of Young David so special.

On Saturdays, we'd gather at the small apartment on Sixty-Eighth street she shared with her parents and little brother. We were far more entranced with her mother, Nora, an artist, than with Miriam. Nora's canvasses lined the cramped dwelling, and we were sure that she was a great painter who had landed in our improbable corner of Brooklyn.

Nora, we decided, was the epitome of cool.

Yet this is what I heard happened when Miriam told her father and bohemian artist-mother that I had cancer:

Miriam was forbidden from seeing me, banned from visiting me in the hospital, told to keep away from me. From my hospital bed, I was told her family believed I was contagious and told their daughter to stay away.

That is how I learned that I was damaged goods.

I had a secret I had to keep deep within me: I could tell no one about the summer of 1973 lest they, too, decide to shun me.

As I sat around with my friends from San Francisco in their dorm rooms, I couldn't bring myself to share my own narrative and the memories that still haunted me, so that with them as with the rest of my peers, I felt as if I were indeed contagious—in quarantine—surrounded by an invisible wall that kept me apart from them.

16

The Spring of Nevermore

Sunday nights were the hardest.

Edith always insisted on accompanying me to Grand Central Station. She would stand there on the platform as I boarded the train to Poughkeepsie—or *Po-Kip-See* as the conductors insisted on calling it in their singsong recitation of all the stops along the Harlem-Hudson line.

Hastings, they'd chant, Ardsley, Ossining, Croton-Harmon, Change at Croton-Harmon.

We'd board another, older train and the singsong refrain resumed: Garrison, Peekskill, Cold Spring, Beacon, and *Po-Kip-See*.

At first, I left New York in the early afternoon. I'd take a seat by the window, finding peace in looking out onto the expanse of the Hudson River. But I dreaded going back so much that I began to delay my return, each week opting for a slightly later and later train. It was already dusk when the train left Grand Central, and the ebbing light only added to my somber mood. From my seat, I could see my mom lingering—she always liked to stay until the last moment, waving and waving to me.

Then I couldn't see her at all.

It was pitch-black when I'd reach Vassar. The train was usually empty except for a handful of other students—rarely anyone I knew. We'd share a taxi to the campus and then go our separate ways. Back in my dorm room, I couldn't shake off the desolation that enveloped me. I'd walk to the central dining hall for supper; it was usually sparsely attended on Sunday nights, as if others had also put off coming back or made other plans.

One Sunday evening I was joined by Betsy, a lovely girl with ash-blond hair who had struck me as friendly and unassuming. Betsy was with a freshman boy I had noticed on campus and liked from afar. He was the son of a college president, he told me, and as he kept talking, I fixated on his sweater, his wonderfully expensive brown woolen sweater that he wore over gray cuffed pants. Both Betsy and he embodied this WASP ideal—and because they were so thoroughly, so completely American in their manner and demeanor, they seemed absolutely foreign to me.

Dinner was congenial and relaxed. I was at last a part of a small group—me, Betsy, and the handsome boy in brown.

I resolved to have what my companions were having: roast beef. I wasn't going to keep kosher anymore, I decided then and there. As it happened, even César, my favorite sounding board in the family, had offered me religious dispensation. During the semester break, my brother and I had discussed in detail how I wasn't eating especially well, how my health was still in jeopardy, and we agreed it made perfect sense to eat the regular fare. I was going to befriend the wonderful WASP boy in the brown sweater, eat whatever I wanted to eat, wear Top-Siders, and throw away any vestige of my old self.

I took a hesitant bite of roast beef, then another, and a couple more. Then, I stopped; as my two new friends talked, I kept pushing the roast beef on the plate around with my fork, not really eating it. Afterward, we walked through campus together, though Betsy left us to go to the library. The boy in the brown sweater kept up a steady stream of chatter, and while I had no idea what he was saying, I realized that I liked his manner, his gait, the sound of his voice, his American accent. I enjoyed simply being with him.

We parted ways in the quad, and as he turned to go to his dorm and I to mine, I wanted to call out after him—WAIT, don't go. I didn't, of course. I couldn't seem to do what many of my peers would probably have managed in a heartbeat—prolong the evening, walk back with him, accompany him to his dorm, his room—which is what I realized I wanted to do.

Back in my room, I felt oddly agitated—angry with myself for having eaten the roast beef, and for not having eaten it. Angry for wanting the boy in the brown sweater and not wanting him enough. Angry that I was playing a new game by old rules. Angry above all that I couldn't change the game or the rules or myself.

I wondered why I had bothered to come back.

There was only one solution to my Sunday night blues. I decided that I would no longer return to Poughkeepsie on Sundays.

It meant missing a couple of my classes Monday, but no matter—I would figure out some way to make up the work. Instead, I began to amble back late in the morning, well after Edith had left for her job at the Brooklyn Public Library. She couldn't accompany me to the train.

And somehow, that made leaving easier.

I'd sit with Leon in the morning as he fixed himself breakfast—typically one modest egg in a pool of melted butter that he prepared in a miniature frying pan, along with a piece of bread and feta cheese. He was always generous, offering to share whatever he was having with me in the same way that months earlier he had fed me his favorite black olives to sustain me during my treatment.

He would be done eating and ensconced in his prayer books by the time I left. I'd try to hug him good-bye. He'd simply nod, as if it didn't bother him that I was leaving again—as if he weren't even mindful of my departure. I would close the door behind me and go.

I arrived on campus in broad daylight, which felt slightly more manageable. The more classes I skipped, the harder it was to keep up or even to focus on what I was studying. The exception was the French Department, which I found more hospitable than any other on campus.

Located in a small building known as Chicago Hall, it had a group of professors I found accessible and friendly. There wasn't a single American in the bunch. Most were from France and had been educated in Europe; at least one was a Holocaust survivor. I happily chatted with them in French, at ease again, as if I were carrying on a conversation with Mom in her kitchen.

As I padded around Chicago Hall, I had a sense of peace that eluded me elsewhere at Vassar.

I was taking a literature class with a Frenchwoman named Elisabeth Arlyck. Professor Arlyck was deeply charming, but she gave tough, demanding assignments. We were reading many of the authors Edith had talked about as I was growing up—Balzac, Flaubert, Stendhal. Mom would always say their names so reverentially.

Nervous about having to analyze and contrast the heroines of Balzac's *Le Lys Dans La Vallée* and Stendhal's *Le Rouge et Le Noir* for a paper the professor had assigned, I decided to seek out my mother's help. That week, I left school even earlier than usual and went home to show Mom my daunting assignment.

I am not sure when Edith had last looked at those novels—probably a quarter of a century earlier or more. I sat by her in the kitchen, and as if she had read the books the day before, she calmly began to dictate an essay about Balzac's shy female protagonist, Madame de Mortsauf, comparing her with Stendhal's Madame de Rênal. The sentences flowed and the thoughts coalesced and the books themselves, which I had found hard to follow, made sense.

When Professor Arlyck handed the paper back to me, I saw that I had earned my first "A" at Vassar—or rather, Mom had. Professor Arlyck scribbled that I had the ability to get straight to the point, "ce qui est une grande qualité."

I had cheated, of course—this was my mother's paper, not mine—but I didn't feel in the least bit guilty. I was simply elated at the thought that Edith, forced to abandon her teaching career at L'École Cattaui at nineteen, could still wow a Vassar professor with her style and erudition.

I was now cutting more and more of my classes.

I'd think, *Why wait till Friday to go home for the weekend? Why not go ahead and leave Thursday night?*

Or even Thursday morning.

I was always looking at train schedules, calculating the next train I could catch. I would race by taxi from Vassar to the ancient redbrick station and sit in the vast, chilly waiting area, staring at the incongruous chandeliers hanging from the ceilings, until they announced the train to New York. Once in New York, I was ready to turn around and go back to Poughkeepsie.

I wasn't spending more than a night or two at Vassar each

week, and even that was becoming an ordeal. I was constantly, frenetically shuttling between Poughkeepsie and New York, New York and Poughkeepsie. No matter where I found myself, I felt I should be somewhere else.

Finally, I made an appointment to see a Vassar psychologist, an older woman who enjoyed a kind of cult following. I walked to her office in Metcalf, a building tucked away in a discreet part of campus, hoping and praying no one would spot me. Seated in front of her, I spoke haltingly about my illness the year before starting Vassar.

To my surprise, the doctor began to cry—and revealed that she herself had recently suffered from breast cancer. I sat there stunned, and oddly more shaken than when I came in. I hadn't expected a psychologist to cry.

Was this an anniversary for me, she wanted to know, turning sober and professional and impassive again.

An anniversary?

The question was jarring—I had always thought of anniversaries as happy occasions. It was February 1974, and I felt very far away from any joyful events. Then I remembered that February night exactly one year earlier when I was getting ready for my friend Celia's wedding and put on my first floor-length gown and noticed my leg was swollen. I recalled the terrible weeks and months that followed as doctors deliberated about what was wrong with me and then diagnosed me and pondered what to do. Celia's wedding—when my life had changed forever.

I nodded: Yes, it was an anniversary, it was the one-year anniversary.

I didn't go back to see the school psychologist. Somehow, her teary outburst had unnerved me. I saw it—unfairly, perhaps—as a sign of weakness or vulnerability, qualities I didn't want in a shrink. I wanted someone tough and strong, someone who could take on my demons and defeat them, someone who could stop me from running. That was what I wanted most of all—I wanted somebody to put a stop to my restless, anguished peregrinations from Poughkeepsie to New York, from New York to Poughkeepsie.

Early one Tuesday morning, I woke up in my dorm room and decided to leave Vassar. Ignoring all of my classes and assignments, I boarded the first train for New York. Once in Grand Central, I wasn't sure what to do with myself. All around me, men in suits and women in high heels were off to work, and they seemed so purposeful.

But where was my sense of purpose?

Months back, when I had confided my distress to my sister Suzette, she had seemed impatient and unsympathetic. "What would you rather be doing?" she had asked crossly, her voice filled with resentment. "Working as a salesgirl at Macy's?"

And because she was my older sister and I trusted her implicitly, I thought those were indeed my only two options in this world—Vassar, or manning a cash register at Macy's—and I'd forced myself to stay in Poughkeepsie.

Now I couldn't—I simply couldn't be *there* anymore.

I decided instead to go shopping. Passover was around the corner and I needed to buy the new dress that had been denied to me the prior year when I was ill. It had always been such a lovely tradition, going shopping with Edith. I had been in the hospital until the eve of the holiday, so a shopping trip had been impossible. Besides, my poor mom seemed beset with worries: Buying me a dress was nowhere on her mind.

Now, I planned to resume my beloved childhood ritual. Though I wasn't a child anymore, and the holiday didn't seem nearly as important as it had once, I was determined to recapture a semblance of the joy I had always felt on the eve of Passover, with me and

Edith venturing out hand in hand to scour the stores for that all-important, perfect holiday outfit to wear at the Shield of Young David.

I headed for B. Altman's on Thirty-Fourth Street and made my way to the sixth floor, home of the Shop for Pappagallos. I sat down and asked an obliging salesperson to bring out several pairs in my size—Pappagallos trimmed with colorful piping, Pappagallos with whimsical tassels, two-toned Pappagallos, Pappagallos shaped like ballerina slippers.

Didn't I deserve to own at least one pair? Surely, surviving nearly a year at Vassar had entitled me to that. With a newfound sense of resolve, I picked out a pair of slender, delicate navy blue flats and paid for them with cash I had painstakingly amassed over months.

I carried my treasure in a small Altman's shopping bag over to the young women's department, where I wiled away the afternoon trying on dresses—beautiful dresses, dresses that brought me back to a time when I liked the way I looked, when I felt so proud about the way I looked, dresses that reminded me of my arrogant years. I slipped on my new Pappagallos to get the full effect, and settled on a short, elegant pale blue empire dress. I was, at last, properly outfitted for Vassar.

Yet now, armed with the accessories that would allow me to fit into my new world, I wanted nothing to do with that world. I was missing a raft of classes and forgoing reading assignments. I was probably risking my scholarship. But none of that seemed to matter on this March afternoon.

Altman's had opened a chic hair salon, and my brother Isaac had been critical of my long hair. I had worn exactly the same style since I was a teenager in Brooklyn, straight and flowing past my shoulder. My brother had made it clear I looked a tad déclassé— as if I had taken Bensonhurst along with me to Poughkeepsie. I stopped to chat with a hairstylist who offered me a brand-new look—one of the fashionably short pageboy haircuts that were all the rage. I climbed into a chair and let her get to work.

As I watched several inches of my hair dropping to the floor, I couldn't help feeling somewhat doleful. It wasn't that long ago that I had despaired as I saw it become thin and fall out after the radiation. It had taken months to grow back, yet here I was cutting it off.

As the stylist trimmed and brushed and blow-dried it into shape, she kept assuring me I would be a new person without my big head of Brooklyn hair. Don't you look great, she kept asking. Don't you look great?

I wasn't so sure, and she seemed cross when I didn't reply. Though my new hairdo was chic, I missed my free-flowing hair.

I returned home to Sixty-Fifth Street with my bundles from Altman's and my semishorn head. Then, I started calling old friends; I was anxious to show off my Vassar look and my Pappagallos—though not to anyone at Vassar.

A few days later, I made my way to a house a couple of blocks away where some of my neighborhood friends had gathered. They all peered curiously at my feet. Rena Douek, who had been close to me in high school, seemed amused: "But what are these?" she said pointing to my shoes. The slender wisplike blue loafers must have looked so odd and decidedly unfashionable.

I had been so eager to leave these friends of mine behind— the world beyond had beckoned with such intensity. And I was also secretly contemptuous of their plans to go to Brooklyn College—the institution of choice for most of my classmates from New Utrecht. Yet Rena and others seemed perfectly content at Brooklyn College—happier and more stable by far than I felt at Vassar. The horror was that I couldn't go back—I couldn't simply admit I'd made a mistake, return to Sixty-Fifth Street, and register at one of the City University colleges. I wasn't at home in Brooklyn anymore, and I wasn't at home in Poughkeepsie; all I could do was frantically, erratically, run from one to the other.

I realized that I no longer fit in with my old friends. I wasn't sure I fit in with anybody.

I was due for my checkup at Memorial Sloan-Kettering. The appointment was my chance to huddle with Dr. Lee, my oncologist. I was anxious and wanted simply to hear that I was fine—the possibility the illness had returned haunted me constantly. But I also viewed it as an opportunity to make inroads with a man I found both elusive and charismatic.

I planned carefully what I would wear, so as to appear confident and, above all, healthy. I wanted to play the part of the mythical Vassar girl he had known in his days at Yale. I settled on my new minidress from B. Altman's.

And, of course, I wore the Pappagallos.

In a rapid yet precise examination, he found nothing out of the ordinary. On the contrary, the fact that I had put on more weight was hopeful. As he poked and probed for whatever mysterious indicators that would signal a problem, he seemed to grow more confident and reassured, and we both relaxed. I hopped off the examining table and took my favorite seat by his desk.

Dr. Lee and I had become close since my malady. I tended to confide in him pretty much whatever was on my mind. But I had no desire to tell him of my despair at college—that going away had turned into a complete debacle.

He was so incisive in the way he questioned me, not simply about my symptoms but about my friends, my classes, even what I was reading, that I often let on more than I'd intended. At some point I blurted out that I wanted "meaning" in my life.

It was the 1970s; everyone was looking for meaning, or talking about looking for meaning.

Everyone except Dr. Burton J. Lee III. He spent his days taking care of the dying, the near dying, and the gravely ill. When he wasn't seeing patients, he worked in Memorial's laboratories, and he had helped develop some of the earliest treatments for lymphoma. Dr. Lee didn't have to think about finding meaning: It coursed through his entire existence.

He didn't strike me as a man of his times. His clothes were elegant in a classical, studiously unfashionable kind of way—a throwback to some 1950s code of WASPdom. Burton Lee was the ultimate master of high prep. He practically breathed it.

Dr. Lee seemed to view the world around him with amusement and contempt. He loved, for example, to wear a tie that had "MCP"—male chauvinist pig—emblazoned all over it complete with embroidered little piglets. It was clear he had little use for the feminists who were among the loudest voices of this loud decade.

"If you want meaning, baby," he said, sounding a trifle impatient, "then go to work for the *Encyclopaedia Britannica*."

He was only joking, of course—he had a devilish sense of humor. But I took him, as always, dead seriously.

I was dismayed that he hadn't made note of my Pappagallos. It was if he hadn't noticed how much I had changed, how different I looked—how I had finally managed to shed my Bensonhurst ways.

Or maybe I hadn't.

That night at home, I told Mom about my checkup—she tended not to go with me to Memorial anymore. From the start, Dr. Lee had made it clear that his relationship was with me, that he didn't want my parents in the examining room. At most, when she accompanied me, she stayed in the large waiting area.

She was overjoyed by the news and insisted on lighting a floating wick in a glass of oil. We didn't have candles at home; besides the custom in Bensonhurst as in Cairo was to use wicks, which my mother would insert in a small juice glass half filled with water, half with Wesson oil. Then, for good measure, she took some incense sticks and lit them. It wasn't exactly the heady *bakhour* of Egypt that she and my grandmother had favored, but it was the closest she could find—tall thin brown sticks that were sold by the bundle in outdoor stalls on Fulton Street. Ever since my illness, she had begun to burn incense nearly every night. "Baal Haness, Amelena Ness," she prayed.

It was an old Egyptian incantation—she was invoking the ancient rabbi Meir Baal Haness, the Maker of Miracles.

The next morning I got up at dawn and caught a train back to Poughkeepsie. I went straight to my room at Noyes and stuffed whatever I could into my suitcase. I put the blue Olivetti in its carrying case and locked the door behind me. I didn't even look around to see what I could be leaving behind.

I knew that I risked failing some classes, but I figured I could request incompletes and get through the semester by mailing in my term papers and settling for any grade my professors saw fit to give me.

My visit with Dr. Lee had at least calmed me down. I felt more hopeful about my health—at least until my next appointment. And that in turn gave me the strength to begin weighing other options. Summer was nearly upon us; I would look for a job, explore the possibility of another college.

Nothing was clear to me except this: I wasn't going back to Vassar again.

17

The Princess of West 116th Street

*I*t felt so odd but here I was, starting college once again. In September, instead of a train to Poughkeepsie, Mom and I took the subway to Broadway and West 116th Street and my new home at Columbia University.

Over the summer I had found a respectable way to maneuver out of Vassar, at least temporarily: I had requested and been granted a one-year leave of absence to attend Columbia as an exchange student.

I had no intention of returning to Poughkeepsie, of course—even thinking about it filled me with dread.

As I'd considered my options in the relative calm of summer—when I had finally stopped running—a leave seemed a less drastic solution than dropping out of college altogether.

I had new luggage I'd purchased with the fifty dollars in Social Security Disability I now received each month in the mail. It was a benefit my family had wangled with the help of Memorial. I didn't stop to consider what it really meant—that in the eyes of the government I was considered disabled. I was thrilled simply to have a steady flow of pocket money.

Fifty dollars seemed like a luxurious amount to me, far more than I'd ever had at my disposal.

Together Edith and I made our way to the dormitory I'd been assigned at 616 West 116th Street. It turned out to be not a dorm at all but a series of suites with private rooms built around a small kitchen.

I immediately felt relieved: I wouldn't have to worry anymore about the humiliations of a central dining hall and whether there'd be kosher food and if I'd be eating alone or in a group. I could prepare my own meals and eat them in my cozy kitchen. "Six Sixteen" as it was called, was for women only, and that, too, was comforting to me.

Edith was overjoyed with my new arrangement. I was in Manhattan. Although I wouldn't be living at home, I would at least be close by. She raved about all the nice girls who were going to be my suitemates and who shook her hand and introduced themselves one by one as we arrived.

She offered to go back to Brooklyn and bring me some food as we hadn't thought of bringing any with us—I had assumed that, like Vassar, there'd be dining halls or student cafeterias.

A couple of my suitemates suggested we walk over to Broadway instead. We spotted a Chock full o'Nuts coffee shop at the corner, went in, and sat side by side at the counter. I ordered a nutted-cheese sandwich because I liked the sound of it—it was their classic bargain lunch fare, a couple of slices of date-nut bread with a layer of fresh cream cheese that cost less than a dollar.

Edith with her gift for slightly mangling names dubbed it *Le Sandwiche Cheese Nutty-Nutty*. It tasted so delicious I decided

I would be absolutely fine at Columbia even if I only dined on Cheese-Nutty-Nutties for the next year.

At a local grocery store that catered to students, we stocked up on supplies. My mother kept adding items to my little cart—bread, milk, butter, potato chips, and a box of Entennman's banana nut cake because she had an almost religious belief in the curative powers of cake.

It seemed remarkable to me to be buying groceries again. Vassar had been so cloistered and remote, its campus thoroughly cut off from town, that even a trip to a supermarket had been an impossibility.

After Edith left, I went to a gathering for new students downstairs. I was determined not to repeat my lonely Vassar experience: I told myself I had to find a way to break through my isolation and reach out to others.

It helped that I felt better than I had in months. Lately, I'd find myself at times wondering if I had imagined it all—the horror of the last couple of years, the fact that I'd had cancer.

No one will ever know, I thought.

I was going to forge a new life on West 116th Street, a life without any reference to the past.

Columbia would be like a foreign country where I would invent a brand-new identity for myself.

A tall, slender girl with honey-blond hair that flowed past her shoulders greeted me with a wave and a broad smile. She was surrounded by other girls and seemed to be holding court: A sea of them had encircled her, and they were all chatting amiably with one another.

She promptly introduced herself as "LaurieLauriefromNorth-Woodmere." She spoke in an anxious, rapid-fire pace. Laurie Wolf was a freshman, and at seventeen, a year younger than me, but we were oddly in the same boat, trying to make our way in unfamiliar terrain.

I quickly learned that Laurie had at least one major advantage—she knew dozens of members of the new freshman class, including several of the residents at 616 who were from the same corner of Long Island. Laurie looked formidable and rich, but unlike the formidable and rich girls I had met the previous year, she struck me as down-to-earth and deeply friendly.

She was also Orthodox, and so it turned out were many of the girls I was meeting.

I was home again, no longer an alien. Although many of the students came from affluent suburban homes in Long Island or New Jersey or Westchester and were in their own way as privileged as the young women of Vassar, they had none of their coldness or standoffishness.

There was significant common ground—primarily the fact that faith and family were as central to their lives as they were to mine. Most of the students I was meeting seemed very traditional, as if the turbulent world of the 1970s had nothing to do with them, even though Columbia was at the epicenter of the turbulence.

It was the fall of 1975, yet for many of the girls, it may as well have been the fall of 1955. I bonded immediately with Laurie, drawn by her effervescence and optimism, which was in such contrast to my dour outlook.

We started having meals together—lunch, then dinner, then lunch as well as dinner, often sitting in her kitchen where we'd be joined by one or two of her suitemates, and I realized I had found myself a real home on West 116th Street.

At the time, Barnard, an old-line women's college, was affiliated with Columbia yet apart from it. It was proud of the fact that it had its own identity, its own professors, and granted its students a Barnard degree. Still, while I was technically enrolled at Barnard, it was Columbia that I loved—the energy of it, the expanse of it, the intense hopefulness of it, the range of classes it offered, the boys that I began meeting in those classes.

My nascent interest in men had vanished after the illness. It was as if desire itself had been obliterated along with the malignancy. But some of those feelings were coming back.

I found to my shock that I was more like my wayward older sister than I'd ever thought possible. I seemed to be drawn to *les blonds aux yeux bleus*—exactly as Suzette had been. And while Columbia had no shortage of Jewish guys, students with similar backgrounds to mine, I invariably developed crushes on boys with sandy-blond hair and last names like "Evans" or "O'Neill" with whom I had almost nothing in common.

I took a walk each day to the English Department where Robert Evans worked as an assistant. He was older—a junior, possibly a senior—with shaggy, dirty blond hair and wide-set blue eyes. I learned fairly quickly that he had a steady girlfriend, but that didn't dissuade me. I kept stopping by to see him and chat hoping against hope for. . . . For what?

I wasn't altogether sure.

I developed a similar passionate attachment to a boy in my French class who was married. I found him wonderfully attractive with dark hair and green eyes and Irish-Catholic charm. I took to leaving after class with him for long walks across campus. He was very kind, but there, too, nothing ever progressed beyond a series of ever more intense conversations.

One of my suitemates gave me a glimpse of how such encounters could work out very differently. She was new to 616 and we'd heard she had recently lost her mother. She began to bring home different boys that she would take to her room and shut the door. They came in shifts so that she received two, four, five guys in a single day, every day.

We were both amused and horrified. One of us finally asked

her about her freewheeling lifestyle. She replied evenly that after the death of her mom, she had decided to deny herself nothing. I couldn't help feeling somewhat envious at her freedom, at the pleasures she allowed herself to have in the midst of her mourning.

I, too, was in mourning, I suppose, mourning what I had been before my illness, mourning my shattered sweet sixteen and my ruptured girlhood and all that was lost inside of me and would never be again. But the difference between us was that I would allow myself nothing.

Part of it was, of course, my background. Old Cairo, Ancient Aleppo, Nouveau-Syrian Brooklyn—none of these places exactly encouraged a woman to be libertine. It was all but unthinkable, even to someone who had fancied herself a rebel, who had longed to shatter dividers. I was as much a prisoner of these antiquated moral codes as anyone I'd grown up with in the women's section.

But it was more than that. My illness had been like a knockout punch; it had extinguished all sense of desire, any belief in romance, and with it, any hope of experiencing one or the other.

I no longer made demands of God to find love or a boyfriend or a husband. I simply wanted to survive.

Yet I was changing. My energy was coming back, and with that so was my vanity.

This was in large part Laurie's influence. She was the archetypal girl from the Five Towns, a wealthy enclave of villages on Long Island with a distinct personality: wealthy, overwhelmingly (and devoutly) Jewish, and a culture that fostered a kind of extreme narcissism. But I found it to be a benign form that went hand in hand with an abundance of sweetness and caring. She and her friends dressed meticulously, agonized far too much about their appearance, and fretted constantly about "socializing." But theirs was a starkly different dress code than the one I'd encoun-

tered at Vassar—worlds apart from the high prep ethos that counseled looking casual, low-key, and almost asexual.

No need for Top-Siders or dark, shapeless crew necks around
here. The mantra was to "make the most of what you have." These
Orthodox girls favored tight-fitting sweaters and clingy polyes-
ter shirts known as "Huckapoos" that were very fashionable and
rather expensive. They added layers of gold chains and rarely ventured outside without some makeup and perfume.

Then there was the obsession with hair. Blow-driers had come
into general use—the bigger, the more powerful, the better. There
were girls in my suite who devoted entire evenings struggling
with the high-wattage contraptions to twist and curl their shapeless strands to create the perfect pageboy. But even more important was finding a great Manhattan hair salon. What had been a
simple neighborhood affair—going to the local beauty parlor—was
passé. Instead, boutique establishments were opening up in the
toniest quarters of the city, promising women not simply a haircut
but a total transformation.

I was definitely in the market for that.

Laurie and I launched exhaustive investigations to find
the leading salons in the city. We favored West Fifty-Seventh
Street off Fifth Avenue, where some of the top hairdressers operated, book-ended by Bergdorf's on one side and Henri Bendel
on the other. The tab typically came to fifty dollars—money I
didn't have.

Then I remembered my Social Security Disability check.

Instead of using it for textbooks or groceries, my monthly
SSI money became my way of paying for voyages to West Fifty-
Seventh Street where I made appointments with the hairdress-
ers favored in the latest issues of *Vogue* or *Glamour*. I'd even call
the magazines' beauty editors to ask for suggestions. Laurie and
I took turns—one month she would go to a fancy salon, then I

Loulou shortly after graduation from Vassar, New York, 1978.

would go some days later. We competed to find the most arrogant, self-assured, outré stylists in New York.

Invariably, I'd find myself standing in front of a young, fashionably dressed male hairdresser who would deliver a withering assessment—my hair was all wrong for my face, it was too long (or else too short), I should never have bangs (or I should *absolutely* have bangs), until I was ready to throw myself at their mercy.

Which I suppose was the point of the torturous exercise.

I would emerge if not transformed then certainly much improved with my sleek new do. I never could maintain the look on my own, but for a couple of days at least, I felt at home among the princesses of West 116th Street, at last almost their equal.

Laurie occasionally invited me to go with her when she left for the weekend. Home was in North Woodmere, a neighborhood of placid ranches and cookie-cutter colonials set on identical one-acre lots. It wasn't exactly luxurious, not that part of it, but it was certainly prosperous—the destination of choice for upwardly mobile families who had sought to flee Brooklyn and Queens in the late 1950s and 1960s. A developer had built scores of starter homes and sold them to couples such as Laurie's parents who longed for a piece of the suburban dream.

I looked around, enchanted by the rows of homes with neat lawns and beautifully apportioned interiors. When we went to visit Laurie's friends, it was often to take part in sample sales. Despite the wealth, hers was a culture consumed with bargain hunting and "buying wholesale." A couple of the adults Laurie knew seemed to have a sideline selling discount designer clothes out of their basements and dens.

I was less interested in the clothes than in the homes. I was struck by how immaculate they were—modern, clean, decorated with magnificent furnishings that seemed to be there almost for show—not really to be used at all.

Each room had furniture sets, not the stray inexpensive pieces that filled our apartment on Sixty-Fifth Street. There were dining room tables with matching credenzas, living rooms where the sofas and armchairs and coffee tables had been carefully coordinated.

Until then, my only glimpse of suburban opulence came from television, so that the living rooms and dining rooms I was entering reminded me of the houses on those shows I'd watched so avidly as a child new to America—or the grand prize on *Let's Make a Deal*, where curtain number two would part to show a pretty model standing in a room filled with wondrous, matching pieces of furniture—the perfect bedroom, the ideal den.

Visiting Laurie was my first exposure to the American Dream—or at least its Orthodox Jewish equivalent. Laurie was the heiress apparent to Philip Roth's Brenda Patimkin, as if Brenda had been transported from Short Hills, New Jersey, and plunked down on the southwestern shore of Long Island. Except that my friend in many ways transcended her background—she

Laurie Wolf, New York, 1970s.

embraced her culture and its values, but was somehow grander than it.

Laurie's parents were originally from Brooklyn, as were many of their neighbors. Laurie herself was born in a one-bedroom apartment on Westminster Road in Flatbush. As Mr. Wolf had prospered, he and his wife bundled two-and-a-half-year-old Laurie and her older brother and left their cramped apartment for the nirvana of grassy lawns and a newly built "hi-ranch."

Yet they didn't entirely sunder their ties with the past. Laurie's dad, Cyrus, an amiable Court Street lawyer who had grown up in Borough Park not far from where I lived, still commuted every day to his law firm in Brooklyn. I got along best with him, and I think it was because I conjured up the hardscrabble life of the old neighborhood. He had been fond of that life, and I think he was fond of me for reminding him of it, reminding him of the days when he was a bundle of drive and ambition and longed to rise far above his background.

Inside Laurie's living room was her mom's collection of Lladro figurines. Mrs. Wolf had amassed dozens and dozens of precious and semiprecious tchotchkes. They, along with pieces of Lalique crystal, seemed to occupy every nook and cranny. No one ever seemed to congregate in the living room. I liked Mrs. Wolf, who was edgy and acerbic, and had a wry sense of humor, but Laurie's relationship with her mother was tense, and the two stayed out of each other's way. My friend and I gathered in the kitchen or den, but I was always anxious as I crossed the living room, terrified I would inadvertently break a delicate Lladro or Lalique and be banished forever from this idyll.

Instead, Laurie and her parents drew me in ever closer to their magical circle.

I had found an improbable city of refuge on West 116th Street and again in North Woodmere. I felt safe, happy, and shielded from my past, even though occasionally I encountered pieces of it.

One of the Five Towns girls at 616 was Karen Alter. I'd noticed her immediately, not simply because she, too, was friendly and expansive, but because of her name, which was the same as the doctor at Maimonides who had first diagnosed my Hodgkin's. I found out very quickly that she was his niece and whenever I ran into her, I was taken back to my last encounter two years earlier with the imposing Brooklyn cancer specialist.

One afternoon, my parents and I had been summoned to his office. He had learned I was seeking treatment at Memorial in Manhattan, but that wasn't his concern. Addressing my mom and dad almost as if I weren't in the room, he made the case for an operation that could enable me to have children later on. There was a simple surgical technique that would shield my ovaries from the radiation. He had heard I was declining the surgery.

"She won't regret it now," Dr. Alter said, his eyes fixed on my mother, "but she will later. The older she gets, the more she will regret her decision. She will regret it more and more with every passing year."

Edith nodded and said nothing. But his stark words must have registered because days later, as I insisted I wouldn't submit to the surgery, she repeated what the doctor had said. She didn't really try to persuade me to change my mind—she was too fearful of modern medicine, and like my father, terrified of hospitals and operations. She simply noted his prediction.

Scared and bewildered, my mother cared only about my survival, as did my dad. There had been so much loss—the baby daughter who'd come along before I was born, our home in Cairo, our very identities. Now, the focus was simply to save me.

My mother didn't offer any counsel about the proposed operation. She didn't advise me about what to do. My siblings were far more assertive. My older brothers calmly suggested I consider it. But from her perch in California, Suzette weighed in most emphatically of all. She didn't think I had cancer, the doctors were all wrong, and I certainly didn't need an operation that could possibly backfire or lead to complications. Surgeries, she told me darkly, often help spread a cancer.

Separated from the family by thousands of miles, yet thoroughly connected to us in her own way—and to me in particular—my sister kept calling to tell me to beware of the physicians, to be careful of Memorial, and to warn me that the surgery would only hurt me in some way she couldn't pinpoint.

Still, a strange whisper of common sense told me I should have the surgery, never mind Mom's fears or Suzette's objections. I made the arrangements and found myself in a drab six-bedded ward in the oldest part of Memorial, a section that had once been a public hospital for the poor. The women were all breast cancer patients who had had mastectomies and assumed that was what I was getting. Terrified, I buried myself in a book—the feminist bestseller du jour, Phyllis Chesler's *Women and Madness*—and tried to ignore them. I requested a transfer to the children's ward, where I'd stayed as a patient at Maimonides. But a young doctor told me darkly, "You wouldn't want to be there." It was even sadder than my breast cancer ward—filled with hopelessly sick toddlers.

On the eve of the surgery, the doctors returned to my bedside with some news. After examining my charts, they'd decided it was prudent to perform a far more extensive operation—so-called exploratory surgery where they'd search for signs that the cancer had spread, as well as removing my spleen lest it, too, was diseased.

My mom and I listened to them horrified. I had expected a relatively simple and brief procedure; now they were talking about a possibly dangerous four-hour operation that would leave me with an enormous scar. The original surgery wasn't mentioned anymore; it seemed suddenly beside the point.

I watched as Mom, shaken out of her passivity, began to cry and to yell at the doctors.

I asked them point-blank if I needed this exploratory surgery: Could I be treated without it? They replied that I could. And that was that. I couldn't bring myself to say to them, stop, can't we go back to the original plan, the original simple surgery? I checked out of Memorial even as the doctors made plans to have me return almost immediately to begin radiation treatments.

I didn't realize the import of my decision, not really. As I marched to my first radiotherapy treatment, I felt mostly numb—it occurred to me that I had made a life-changing decision, a choice that would change me forever, but I didn't know what to do about it other than keep walking to the treatment room.

The regrets had begun only later, after I'd finished the therapy and was ensconced at Vassar and now, at Columbia. I would find myself haunted by what Dr. Alter had said, the coldness with which he'd said it.

I'd think about his words whenever I ran into his cheerful niece, and the longing to run would overtake me. But then I'd meet Laurie for dinner, or take a walk across campus and visit Robert Evans at the English Department, and I would calm down, forget all over again.

Over the years I would always be asking doctors the same question: Should I have had the operation? Would it possibly have worked? I secretly hoped that they'd say no, but I learned that the smaller surgery could indeed have been effective. As for the larger operation they had pushed, it fell into disfavor. It turned out to do more harm than good, and many patients who were subjected to it developed awful complications and died.

One night at Barnard, I dreamed that I asked César for a bottle of Joy; my older brother spoiled me—even more so since my illness. I thought the perfume by Patou was the ultimate status symbol, and I longed for it in real life. But in the dream, I was very upset because I didn't get Joy after all. Instead, César pulled out a bottle of Cache, the French perfume whose name means "Hidden."

*W*hile visiting North Woodmere one weekend, Laurie decided we should stop at a designer shoe outlet. To my surprise I spotted that old object of my desire, Pappagallos, but at drastically marked-down prices, and I found myself telling my friend about my misadventures and obsession with the shoes, how they had overshadowed my entire Vassar experience. Laurie listened with her usual skepticism and tried several on. Without pondering the Great Meaning of Pappagallos, she promptly bought a pair.

They were pretty and elegant, but they looked like other shoes—

not at all the pastel flats I'd seen at Altman's. Even Pappagallo was changing, producing shoes that resembled the popular styles women favored, and not a throwback to some idealized, vanished, 1950s WASP America.

Late in the fall, it seemed natural to join the Wolf family for Thanksgiving dinner. They would be observing the holiday at the home of Laurie's Aunt Eleanore in Hewlett Harbor, a community whose name my friend pronounced in hushed tones, urging me to prepare myself for the ultimate in luxury, beauty, and excess.

I was afraid to tell her that Thanksgiving was a strictly theoretical holiday in my mind. We had never really observed it at home even though we had now been in America more than a decade, and I had recently become a U.S. citizen. Turkey was foreign to us, not at all part of our Levantine fare. But it looked delectable in the pages of the magazines I leafed through, which invariably featured recipes and pointers for the "perfect" holiday table.

In our first years here, Thanksgiving had been virtually unknown among the families of the Shield of Young David. Our lives revolved around the religious holidays and only the religious holidays. As we became more Americanized, turkey began to make a sneak appearance in some homes. My friends, the Cohen sisters, decided to prepare it for the *Shabbat* lunch on the Saturday immediately after the holiday. They would invite me to join them and while I relished the delicious spread, it felt more as if we were enjoying a grand Sabbath feast than observing a secular American national holiday.

Edith tried to satisfy my longing for "*∂u* turkey." At our local Key Food, the small supermarket on Twentieth Avenue, Edith had found a cooked kosher turkey breast in the frozen foods section. It came in a tinfoil container and had an enticing image of a golden brown turkey on its wrapping. My mother plunked down an exorbitant amount for it and then tried to follow the elaborate directions on how to prepare it. Yet no matter how carefully she cooked it—and how excited I was at the prospect of eating turkey—the final product was inedible. Yet my mother continued to buy it year after year, eager to have me enjoy a semblance of this holiday that filled me with so much longing.

Thanksgiving in Hewlett Harbor offered me a taste of that elusive all-American suburban life I craved. In a magnificent home where paintings covered every wall—Laurie's aunt was an art lover—we were seated at a long table whose centerpiece was a massive roasted turkey.

It tasted every bit as wonderful as I'd imagined, and not a bit like the turkey breast Edith had insisted on making the day before, though I had assured her she needn't bother.

As I left, Mom seemed both happy for me but also rather wistful. Though she loved this new world I was entering, and understood that it offered me so many more possibilities than she could provide, she still wanted to compete with that world even when it was a hopeless competition, even when she was clearly waging a lost battle against a Hewlett Harbor turkey.

As the new winter semester started, I enrolled in a playwriting seminar offered at Columbia. I had gone to college hoping to follow Mary McCarthy's guiding star, yet I hadn't taken a single writing course. Vassar's English Department had been so thoroughly alienating, I had retreated to French and other subjects instead.

In a small room, about a dozen of us sat around a horseshoe-shaped table. Each week, we'd read out loud pages of dialogue to the professor and our classmates and learn from their feedback. There were no requirements, no exams, no assignments.

We were supposed to do nothing but work on plays.

Most appealing of all was the makeup of the class—all male except for me. A soulful-looking senior named Sean O'Neill was accorded a seat of honor near the head of the table, close to the professor, who treated him as if he were Eugene's heir and descendant.

When Sean O'Neill read, we all listened spellbound. He had a wonderful manner about him, though I was perhaps as riveted by his hair, which was soft and blond and fell over his eyes, as by his lines of dialogue. I was once again excited about my classes—I couldn't wait to get to my playwriting seminar each week and hear Sean O'Neill read. But I never seemed to be able to strike up so much as a simple conversation with him.

Though he was always amiable enough, and smiled at me, he seemed beyond my reach.

Back home at 616, I would tell Laurie of my hopeless crush on this boy. She couldn't advise me; in the same way that she had arrived at Columbia with a coterie of girlfriends from the Five Towns, she knew many of the boys from her area. They had grown up together, gone to the same schools and sleepaway camps, or else they were close to her older brother, Eddie, a junior at Columbia. If I'd been drawn to one of them, it would have been easy. Using her extensive social network, Laurie could have provided chapter and verse on the boy's family, prospects, background, his reputation back in high school, and whether he was a worthy object of my affections. She would have done what she could to help the romance, since in her world, there was a premium placed on matchmaking and "fixing up" people.

And yet even as I kept falling in and out of love—from afar, of course—it was never with one of the perfectly nice and presumably attainable Orthodox Jewish boys who would have been more compatible with me on a hundred different levels, from our shared religious backgrounds to the holidays we observed and the generally conservative cultures that had formed us.

These boys had one significant trait in common. While they were coming of age in a period marked by a push for social freedom, they had the values of a much older era. From the moment they arrived at Columbia, they were—every bit as much as Laurie and her girlfriends from the Five Towns—on the lookout for a mate, the person with whom they'd settle down, start a family, and share the fruits of the spectacular achievements they had every intention of attaining. They were driven and ambitious, intent on going to medical school or law school or business school—nothing else would do. But unlike other driven, ambitious young American men, the notion of settling down at an early age was deeply ingrained and even sacrosanct.

I knew that I would never pass muster.

At the first hint of a serious entanglement, I would be obliged to tell all, to reveal my sad, sordid story, and what would happen then? To my mind, the mythical Sean O'Neill as well as the yarmulke-wearing Jewish boy from the Five Towns or the outer boroughs would both leave.

As I wandered through that great campus bazaar, the *souk* of love and romance that flourished at Columbia and every other college campus, where men and women came together to barter for social and sexual favors with all the wiles of traders in an old Middle Eastern market, I stayed resolutely alone, persuaded I had lost most of my value.

It didn't matter that my new world was more enlightened than the one I had left behind. It didn't matter that I had found myself a home where education was valued. It didn't matter that here, a woman was encouraged to study and flourish and become a professional. It didn't matter because inside of me, there was still the tune that had haunted me at Vassar. It was a variation of Helen Reddy's anthem to the women's movement, *I Am Woman*, which I had first heard in high school. Back then, I had made it my own, embracing all it signified. The song had followed me to Poughkeepsie and now to West 116th Street.

Except it was a more personal rendition I kept hearing now, a more anguished refrain: *I am not woman enough.*

S till, I was changing or maybe simply changing back into the person I had once been.

I found pleasure in my studies, and some of my old arrogance was creeping back into my demeanor. I enjoyed surveying the world around me from my regal perch. I was the Egyptian princess again—albeit with a tarnished, broken tiara. After a year wandering around with my head bowed, wearing dark clothes, hoping no one would notice me, I wanted to be noticed again.

Which is, of course, what happened as I tooled across the campus and the city with Laurie.

She was very charismatic and charming, an expert at attracting attention, and wherever we went—to Central Park, to our favorite Broadway eatery, for coffee at the Hungarian Pastry Shop, or uptown to the Cloisters—we were often being approached, and men seemed eager to get to know us and the city was ours.

With my newfound confidence, some of the old bravado also returned.

I enjoyed regaling Laurie with stories about my family's illustrious past—our grand life in Cairo. But then my mother would come to 616 to hand-deliver the small hamburger patties she insisted on preparing for me each week and the banana nut cake from Entennman's she thought was essential to my survival, and Laurie couldn't help noticing how broken-down Edith seemed— how different from the self-assured and pampered mothers she knew from the Five Towns, women whose lives were crammed with trips to beauty parlors, who would never be seen leaving the house without their nails polished and their fingers bejeweled and their hair curled and sprayed and lacquered into shape.

Edith, she noticed, never had her hair done; it was loose and gray and slightly disheveled. She didn't wear a stitch of makeup. Did she ever have her nails done? Did she not own any jewelry? My friend had trouble reconciling the grandiose stories I told of my family's past—my own exotic princess airs—and my impossibly humble, self-effacing mom.

Edith was rather shy around my Barnard friends. This was a departure from the way she had been when I was a child in Brooklyn: She had loved to banter with anyone I brought home and engaged my companions in conversation. But that had all changed, and it was as if she had suffered her own spectacular loss of arrogance and had trouble interacting with others.

While I was making a comeback, my mother was still wallowing in the overwhelming despair of 1973. She was, in her own way, as broken down as I was—perhaps more so, because I am not sure she ever really recovered. Since she could never speak to me openly about the illness, could never even allude to the fact that I'd never have a child, she suffered alone and in silence about the tragedy of *pauvre Loulou.* Only occasionally did she bemoan out loud *le malheur*—the bad luck—that had struck the family.

*C*ome spring, I made a shocking discovery. I learned that the prior year, several Columbia students had been arrested for drug trafficking, including allegedly selling cocaine. One of those arrested was Sean O'Neill.

The chasm between us was immense—far greater than I'd realized. I had been able to shut out the more jarring aspects of the popular culture by sticking to the sheltered world of 616. I could pretend there were no drugs at Barnard, or that my promiscuous roommate was an anomaly. But learning of the arrest brought the world of the 1970s into high relief once again.

And along with the reminder that the world around me had changed, that it was evolving at breakneck speed, was the sense that I wasn't a part of that world, for better and for worse.

One evening, Laurie and I were loudly bemoaning the fact that neither of us had ever smoked—not even a cigarette let alone mari-juana—and were clearly out of step with most of our peers. One of her suitemates, a worldly senior, overheard us and invited us to her room. She casually lit a joint and passed it around.

Laurie and I were giggling nervously. We felt as if we were about to commit the greatest sin imaginable, and to us I suppose it was. And neither of us had the courage to do much more than place the joint in our mouth before handing it back to Laurie's roommate.

We didn't inhale and we didn't exhale. But we felt awfully jolly afterward, happily asserting we had "tried" pot, as if that made us honest-to-God, true-blue members of our generation.

I had a major decision to make: Vassar had sent forms for me to fill out for the following school year. They had taken me at my word: I would be returning along with other members of the Class of '77 who were in exchange programs at other colleges or studying abroad. I had to sign papers, commit to reenrolling, and choose a dorm.

"You'd be crazy to go back," Laurie declared with her usual bluntness.

A part of me agreed with her, of course. What could be cra-zier than returning to a place I had fled in such despair? But I felt compelled, like the criminal (or victim) who is obsessed with returning to the scene of the crime. Mom, of course, was wholeheartedly in Laurie's camp. She was praying that I'd stay at West 116th Street and abandon this grand illusion of being a Vassar girl.

Because that is all it was, of course, an illusion. It wasn't my identity and never could be.

I wasn't entirely sure what was propelling me back to a place I'd hated. At some level I couldn't let go of my idealized notions of Vassar, my sense of it as a passport to a dazzling future. I also needed to prove to myself that I wouldn't run away again.

My last days at Columbia, I paid a call on Robert Evans, my friend from the English Department. He suggested to my surprise that we go share a bottle of wine. Off we went to a liquor shop on Broadway, where he peered intently at the bottles on the different shelves.

"Do you like May wine?" he asked, reaching for a bottle from the store refrigerator.

I nodded emphatically yes, though I'd never tasted May wine. I loved the sound of it, I loved the idea of it.

We walked back to his apartment and drank the bottle. It was a German vintage, and I found it sweet and utterly delicious. We chatted and then parted amiably. If I had hoped for a grand seduction—or if he had planned one—it wasn't meant to be, and, besides, I was terrified of it.

Instead of moving back to Brooklyn, I rented a room in a Columbia dorm for the summer. I had made up my mind. I felt infinitely stronger and calmer than I had the prior year. I was determined to return to Poughkeepsie if only to prove that I could return, if only to prove to myself that Vassar wasn't going to defeat me ever again.

BOOK FOUR

The Lady
in the
Pink Bow

THE UPPER EAST SIDE:

1987–1994

· 18 ·

The Lost Art
of Penmanship

My older brother César is a man of faith, devoted to God and prayer. Yet this is what he believes happened to our family: that a cruel and vengeful Lord set about punishing each of us for our arrogance in the most painful manner imaginable, so that it was as if a curse—a curse of biblical proportions—had been placed on our household.

Our father, Leon, who strode like a colossus through the streets of Cairo in his white sharkskin suits, towering over everyone he met, never recovered from his fall and found it increasingly painful to walk; at the end, confined to a wheelchair, he couldn't even stand up and the world towered over him. Grandmother Alexandra, so helpless—dependent on us for her every breath—ended her days alone in a foreign land surrounded by strangers. Suzette, my older sister who had spurned us in her youth and scoffed at the very notion of hearth and home, who had moved farther and farther away from the family—Queens, Miami, L.A., San Francisco, London—finally returned. It was too late, of course; both Leon and Edith were gone. She had a child, a boy, Sasha, who became the doctor that she had longed to be, yet lived hundreds of miles from her, and in her solitude, I suspect she finally understood the years when Mom would plead with her to come back. César himself, the only whole one among us, the keeper of the flame, married a wonderful woman, a teacher, and had a family, two lovely daughters, but never the son he craved. There would be no one to carry on our vaunted name, he lamented.

Then there was Mom of the luminous intellect and eloquent prose and graceful penmanship. After Edith suffered the first of her many strokes, her mind simply crumbled. Her once-elegant sentences became choppy and incoherent.

And she knew it. That was perhaps the worst of it—that at sixty-four, she was able to observe her own downfall, to chronicle it, remark on it, and despair over it. She cataloged her losses and afflictions as thoroughly as she had cataloged thousands of books over nearly twenty years at the Brooklyn Public Library.

In the fall of 1987, my mother took over one of my reporter's notebooks that she had found lying around the house on Sixty-Fifth Street. It was a used pad filled with lots of notes from my various assignments. I was working as a journalist in New York, and notepads were my most essential possession. I had returned to Vassar after my year at Columbia, and though it never became Mary McCarthy's Vassar, I did finally manage to graduate. I went to work as a reporter, first for the *Brooklyn Spectator,* a small weekly, and then for Jack Anderson, a nationally syndicated columnist whose articles appeared in the *Washington Post* and hundreds of other papers. Anderson was regarded by many as the foremost investigative reporter in the country, an avenger of the downtrodden. His voice was that of barely controlled outrage. I had to quickly master the lingo: A State Department diplomat was a "Foggy Bottom cookie pusher," the U.S. government was "Uncle Sugar," and any Latin American dictator was routinely called "a tinhorn dictator of a banana republic."

I was thrilled to be one of his "junior muckrakers." Assignments were exotic and often impossible. Mine was to cover the State Department and ferret out secret cables, a mission I undertook with zeal, cultivating diplomats who were loath to leak me any documents, let alone the classified ones Anderson cherished. Later, he sent me to South America to hunt for Dr. Josef Mengele, the infamous "Angel of Death" of Auschwitz.

"I would like you to penetrate Nazi circles in South America," Anderson told me in that hypnotic way he had, and off I flew to Asunción. I hadn't a clue how to look for a Nazi war criminal, let alone one who had eluded capture since the end of World War II. There I was in my twenties, with no experience as a foreign correspondent (let alone as a Nazi hunter), and only one semester of Vassar Spanish, wandering around Paraguay asking total strangers, "Donde esta el Señor Mengele?"

I didn't find Mengele in Paraguay. But in trips to Israel and across America, I did locate some of his youngest victims, twins who had been subjected to his horrific medical experiments at Auschwitz when they were children and had never spoken of their past. My articles about the infamous doctor and his long-lost twins helped spur the U.S. government to launch their first major manhunt. As the world took notice of a war criminal they had all but ignored for decades, I felt as if I were living up to the ideal of Emma Peel—I felt like an Avenger.

I'd returned to New York when Edith fell ill in 1987, and now worked at the *New York Post*, my first major daily. Assigned to the city desk, I started covering breaking news—a murder in the Bronx of a mother and her little boy; the disappearance of a disabled girl in Staten Island; the death of Rita Hayworth on Central Park West.

After my shift ended at night, I came home to Mom. I'd often find her bent over one of my notepads. She would be scribbling over old notes or filling pages that still had some lines left. In her own halting way, she struggled to confront the effects of *la strocke* as she called it, a strange term entirely of her own invention, not quite French and not quite English, misspelled in two languages. In regular entries, she tried to master her unwieldy thoughts and recapture the words that had once flowed from her so effortlessly that she could stand at the stove and dictate a paper I needed for my college literature class while fixing herself a *café turc*.

To my eyes the most disturbing evidence of Mom's decline was in her penmanship. Once the epitome of elegance and style, it had become as crazed and helter-skelter as her mind. The words and sentences that had been so clear and lucid now appeared in strange shapes and odd sizes. They were too big or too small, they'd run upward or downward on the page or else in zigzag, as if scribbled by someone who can't control their pen or their mind.

I suppose that she wanted to restore some order by putting her thoughts to paper. Or perhaps she needed to prove to herself that she still had thoughts she could put to paper. The effort—and it was Herculean—made Mom painstakingly methodical.

Every day, and often several times a day, my mother would begin by posting an entry with the date and hour. She was obsessed with tracking her medications, *les pillules*—the pills—as she called them. She'd carefully indicate that she had taken her various medicines and was following the doctor's orders. She was trying to be a good patient.

It was too late, of course. There had been so many years of self-neglect and punitive behavior. There were the years when she didn't take her pills with any consistency and her blood pressure soared out of control. There were the years when she was always running, running. There were the years of constant fasting for this religious holiday, that holiday. And then came the years when fasting became a way of life. We had all left home, and she simply stopped eating. For a meal, she'd boil herself an egg and sip *café turc* while seated in Dad's old armchair and wonder where we had all gone.

Yet she shrugged off any and all signs that she wasn't well. At most, she would go see her old internist at his office on Twentieth Avenue. You could never make an appointment to see this neighborhood doctor—he simply posted his hours and people arrived and waited their turn, sometimes for an entire afternoon. They'd be seen one after the other, five minutes per person, and then they had to pony up some cash for the visit, which they handed to the doctor's wife, who always seemed in a crabby mood, as if she resented the lot of us. Though the doctor was amiable enough, he rarely offered any thoughtful consultation, or performed an examination beyond the cursory use of his stethoscope.

The little water pills he prescribed didn't come close to doing the job, and besides, she didn't even take them. I wondered if he ever impressed on my mother how sick she really was—none of the assembly-line doctors who took her ten-dollar bills did.

The hearth had never been rebuilt. The hearth was gone, destroyed, eviscerated, and I couldn't even remember the last time she'd said, "Loulou, il faut reconstruire le foyer," which she'd made the mantra of my childhood. It would have meant she still had hope in her heart, when in fact all hope was gone.

Those months that preceded her stroke, she ran frenetically and all the time. In the morning she ran to work at the Brooklyn Public Library, taking one train and then another. She was so frightened of being late. Her new boss, Dallas Shawkey, had instituted a reign of terror since taking over and systematically destroyed the Catalog Department's collegial culture. The job that she had loved with all her heart turned into a nightmare.

If she was late, even by a few minutes, she was penalized. If she made an error, or seemed not to focus, Shawkey took her immediately to task. Finally, she was forced out of her beloved Catalog Department and transferred to a distant branch in Brooklyn where her job was in production—grunt work—cutting and pasting book jackets. She had developed severe arthritis by then, and it was painful merely handling those big scissors. She had also begun to lose her coordination so that her hands shook and were unsteady.

She would try and try again but it was no use. She got into more trouble for not doing her work properly.

A colleague, Devika, was concerned by how fragile Edith looked of late. She was always bent over—she didn't straighten up even when she walked. Devika also noticed that my mother was always rushing to leave at the end of the day, racing out the door. Why the hurry, Edith? Devika asked. Mom broke down and told her of her sick husband who needed her constant care and attention, a husband who was much older than her and was bedridden.

What she didn't let on was that Dad was now in a nursing home. After a bitter fight within the family, Isaac had moved to put him away in a facility in Brooklyn. My father was distraught from the start. It was all for Mom's sake, Isaac kept saying. Dad had gotten to be a handful as he grew older, a big man with lots of physical problems my mother couldn't handle. Unfortunately, life didn't get a bit easier for her after he was committed. In a way it became harder. Every afternoon, she bolted from her job, anxious to get back to the empty house on Sixty-Fifth Street despite the fact that no one was there.

Suzette, the first to leave, stayed true to her word and didn't live with us again. Isaac had also never returned after college.

And I went away to Washington and threw myself into my work for Jack Anderson. César, the most faithful and true one among us, the one who remained with our parents the longest, moved away to be with his new wife in Queens.

I suppose that he saw that as a step up. There was a time all of us thought anywhere on earth was a step up from Sixty-Fifth Street.

With Dad in the nursing home, Edith found herself all alone. There she was in that careworn apartment filled with ragtag pieces of furniture and all those books on the shelves—his prayer books, her French novels. After a long day at work, she'd rest in his old armchair by the window and nibble on the one or two breadsticks that constituted her entire dinner. Near her, on an end table, were photos of me at different stages—framed pictures of me as a child, at Vassar, as a young reporter. She would sit there staring at them, sipping her beloved *café turc*.

She couldn't stay for long in that empty apartment without feeling anxious. And so she'd force herself up from the chair and run to see Dad at the nursing home where Isaac had placed him. That meant taking a couple of more trains until finally getting out at an elevated station, which required her to climb up and down steep flights of stairs even though by this time of day, she barely had any strength left.

Somehow, she would find her way to his room, sit by his side, and coax him to eat, because, unlike her, Leon still enjoyed food. She'd offer him a peach, a banana, an orange, some *mesh-mesh* (apricot)—whatever she'd stuffed in the little shopping bag she carried at all times. When visiting hours were over, and she'd stayed with him as late as she could humanly manage, she'd run back to the subway—always spurning our offers to call for a taxi—for the exhausting train ride home.

She ran so much and so fast that she finally collapsed and on

Leon, in a wheelchair, with a fragile-looking Edith standing nearby, Brooklyn, 1986.

a balmy night in March 1987, a policeman found her lying on the steps of the Fort Hamilton Parkway subway stop a couple of blocks from the nursing home.

I was living in Washington, D.C., at the time, but Mom and I had a habit of chatting every night after one of my own grueling days at the office.

I was anxious when I didn't hear from her. Finally, at midnight, César reached me. Mom had suffered a stroke. She was still alive—barely—in the intensive care unit of Maimonides Hospital. It was too late for me to get back to New York—there were no trains, no planes, no buses, and besides, I didn't even have any money on me. Come dawn, I knocked on the door of a kindly neighbor and asked her if I could borrow enough to get home. There were few ATMs at the time, and I barely had any savings. Armed with my neighbor's cash, I caught the first shuttle flight home.

When I reached the hospital, I went immediately to the private room where Mom had been transferred. She was unconscious: She had sustained a massive brain hemorrhage, and there was so much swelling in her head that she was in imminent danger. I was told that only if she were kept stable and the swelling gradually reduced would she survive.

She seemed in agony those first days, constantly stirring in her bed, oblivious to all of us, aware only of her own torment. If God had shown any mercy toward my mother, it was in this way: She was blessed with extraordinarily compassionate doctors. One, an Orthodox neurologist named Dr. Keilson, watched over Edith as if it were a religious mission. I turned to him constantly. A resident, Dr. Nutti, was so deeply caring he hovered by her side day and night. Dr. Nutti was older and more mature than the typical resident, and he was also European—a native of Italy transplanted to Brooklyn.

Was that what made him more human? I wondered. My impression of hospital residents was that they were a young, rather impersonal bunch. Their lives were all about rotations, grand rounds, subspecialties. There was never room for the personal— they knew they'd be moving on to their next patient, their next ward, their next mandate on the road to becoming a full-fledged physician.

Dr. Nutti was different; to him, Mom's illness was intensely personal.

Edith would be back "at 80 percent," he told me one night when he saw me looking especially forlorn.

Well, I thought, 80 percent of Edith was still more than 100 percent of most people I knew on this earth.

I lingered that night by her bedside, as I liked to do. There didn't seem to be any purpose in going home. The house felt terrifyingly empty. The rooms were in various states of disarray. By the end my mother—so tidy and meticulous by nature—had been too tired and overwhelmed and melancholy to maintain even a modicum of order.

One Sunday, I took a break from the hospital visits and went to see Laurie, my old friend from Columbia. We princesses of West 116th Street were in touch again, and she'd invited me to her grand new house on Long Island, in the leafy Five Towns where she'd grown up.

She greeted me, as sunny and ebullient as ever, at the door of the imposing home that she and her husband had recently bought. All around her, small children were playing, hers as well as her sister Trudy's. The two were neighbors, and their parents were still nearby in the old house, so that the original family had been miraculously preserved, even as Laurie and her siblings grew up, got married, and raised families of their own. We drove to nearby Lido Beach and spent the day together laughing and reminiscing as her three children—miniature blond versions of Laurie—played nearby. We'd be interrupted by Lani, Laurie's demanding four-year-old who seemed to drive my friend to distraction. Laurie would alternately chide her, fuss over her, put an outsized bow in her hair, hand her a beach toy, and we'd resume our conversation only to have Lani return to us, clamoring for more attention.

Years earlier, during another nadir in my life, Laurie had managed to be extraordinarily affirming. That was partially why I'd decided to make the trip, to take a break from Mom and the hospital and the doctors. I prayed that my friend would once again work her miracles on me.

And yet when I came home that night, I felt sadder then ever. My old room felt suddenly so small, as if I had grown but it had failed to grow with me. I understood what my mom had faced on Sixty-Fifth Street, why she was always running.

What, I wondered, did my family have to show for all our years in this country?

After a couple of months, Edith was finally ready to leave Maimonides. She needed rehabilitation, we were told, but where?

No one wanted her. It was my first glimpse of how our medical system really worked in the 1980s, how it tended to see patients not simply as ailing or well, but as desirable or undesirable. Contrary to their do-gooder image, hospitals had become intensely mercenary, focused primarily on their bottom line. By 1987, rehabilitation was considered almost a luxury; it was being rationed, allotted only to the lucky few who could show rapid, definitive improvements. If a facility felt it could get abundant reimbursements, they welcomed a person with open arms. But Edith, a Medicaid patient with a deeply uncertain prognosis, was perceived as risky and even worse, unprofitable.

I appealed to one major facility after another to take Mom. One after the other turned us down: They didn't think she would improve and weren't willing to take a chance. One day, I traveled to the Rusk Institute—the Holy Grail of rehab facilities—to make a personal plea to one of the top administrators. I all but begged him to take Mom, all to no avail.

At last, I was able to get her to a brand-new department that had opened at Mount Sinai Hospital on the Upper East Side. Its director seemed friendly and amenable to taking Mom in. Sinai's unit wasn't too selective yet—the priority was filling beds. Within a couple of days, Edith was transferred to a cheery, sun-drenched ward that required patients to wear "real" clothes as opposed to hospital gowns. I was told to purchase some loose-fitting trousers and blouses for her so that her exercise program and recovery could begin.

Life seemed stunningly hopeful again. I went on a shopping spree at Berta Bargain Store on Eighteenth Avenue and bought dozens of stretch pants and T-shirts and a pair of white lace-up sneakers and remembered how happy Mom and I had been once upon a time foraging through Berta's bins for sale items.

Maybe Edith really will be back at 80 percent, I thought, as I picked out a colorful cotton blouse for her.

Within days the Sinai doctors were demanding to meet with me. The news was grim: They didn't see any point in continuing the rehabilitation. They used the dreaded "A" word. Mom had a form of Alzheimer's, they declared, and wouldn't benefit from the regimen of therapies the unit could offer.

"But you have barely started," I argued. I was still reeling from the stroke and its aftermath; now I had to contend with one of the most hopeless diagnoses a person can receive. And yet when I'd visited her that morning, Mom had looked a bit like her old self in her canary yellow pants and jaunty white lace-up sneakers. She seemed thoroughly enchanted with her sneakers. She had never before owned a pair.

The doctors kept shaking their heads. Edith couldn't grasp even basic commands, they said. She would only get worse. The only route for her would be a nursing home. She could join her husband, a social worker chimed in helpfully, as if she were suggesting a romantic reunion in some exotic locale and not life for two in a bleak institution in Brooklyn.

They wanted to immediately start the paperwork to have her transferred.

I flew to Washington to liquidate my apartment, and also to clear my head. I hadn't had a chance to go back since Mom's stroke. I couldn't stop crying the entire plane ride. An amiable gentleman seated next to me turned out to be a physician in New York. I told him of what had happened with my mom, and he urged me to go see a friend of his, a neurologist named Sidney Diamond, who worked at Sinai.

Before I could do that, I needed to deal with my apartment. I'd put an ad in the *Washington Post* advertising a moving sale. A man arrived early in the morning and offered me a hundred dollars for all I had acquired in my years there—my sofa, my bed, my dining room table, my chairs, my posters, my clothes. I was glad to be rid of them; none held any meaning for me anymore, and I wanted no reminders. I took the money, let him remove every single one of my belongings, and flew back to New York.

Once back at Sinai, I found Dr. Diamond in a small basement office of Sinai. The elderly neurologist with a bow tie and a courtly manner listened intently as I described my struggles with the rehabilitation unit. A curmudgeon by nature, Dr. Diamond relished a war with the hospital bureaucrats who had sidelined him years earlier.

"Make them do it," he replied. "Make them rehabilitate Edith."

Dr. Diamond wasn't sure if Mom was capable of improving, but he passionately believed she deserved a shot. Any kind of therapy—physical, speech—could only help her, he said again and again. The exercises would improve her state of mind—she was surely depressed from the effects of the stroke—and help her walk and move her arms and legs, atrophied from weeks of lying in bed. Besides, simply interacting with therapists and aides would alleviate the terrible isolation he suspected she was feeling.

Egged on by Dr. Diamond, my battle with Sinai began in earnest. One by one, I'd corner doctors and physical therapists and plead with them to keep working with my mother. They explained again and again that her memory was gone, that she had no comprehension of what was going on around her—why, she didn't even know where she was, they argued. For every additional day she was allowed to remain in the rehab unit, they acted as if they were doing me a favor, as if this were all for my benefit.

And yet remarkably and counter to their predictions, Mom was beginning to make progress. I would arrive in the morning and find her walking—unsteadily—with a physical therapist. She occasionally even smiled, a slight, crooked half-smile. Her speech was beginning to come back—she spoke almost in complete sentences, though they were garbled and tentative.

She wanted to go home, she told me one morning. She wanted to return to the house on Sixty-Fifth Street.

She mouthed the word *home* with difficulty.

This is how Sinai's doctors fought their war with me: by attacking any illusions I had left, by launching search and destroy missions to eradicate any and all hope that Mom would make a comeback.

Every day, they would approach her with a faux friendliness that, in her innocence, she mistook for caring. "Edith, how are we this morning?," they'd say, in voices that radiated cheer and bonhomie. She would look at them with her big porcelain doll's eyes, at first somewhat bewildered, unsure of what these men in white coats wanted, but then she would smile, trying even now, in the depths of her devastation, to charm them.

She may have lost her memory and her ability to walk and talk and think clearly, but she could still be utterly lovable.

The pleasantries over, the examination would begin. A doctor would proceed to ask Mom a series of questions that seemed innocuous at first—the types of questions a child would be able to answer. This was a standard neurological test, designed to measure a patient's "cognitive functions"—to ascertain their level of awareness, their memory, their sense of time and place. It was supposed to be scientific, but I came to dub this process "Quiz Show," because it was in fact a sham, as crooked and corrupt in its own

way as the game shows of the 1950s that were contrived to produce a clear winner and loser.

In Quiz Show as it was played at Mount Sinai, the doctors
always had to win—and Mom always had to lose.

A physician would begin by asking "What day of the week is
it, Edith?" He would proceed to question her on what year it was,
what month, what season. Invariably, he wanted to know who was
president—that seemed terribly important, getting her to identify
who was in the White House: What is the president's name, Edith?
The doctor would press her, throw in some choices either to help
her out or throw her off—I was never sure. Is it John F. Kennedy,
Edith? Jimmy Carter? Ronald Reagan?

Did she realize that she was in the hospital? Did she remember
what he'd told her the previous morning, during another episode
of Quiz Show—that she was a patient at Mount Sinai Medical
Center on Fifth Avenue? Where are you now, Edith? Do you
know you are in a hospital? What hospital, Edith? What's the hospital's *name*, Edith?

My mother, my dazzling intellectual of a mother who had read
all of Proust by the age of fifteen, who had accepted the key to the
pasha's library, would look up at the doctor, her doll's eyes growing ever wider, showing fear, then embarrassment, and then abject
sorrow.

She didn't know the answers. She couldn't reply to a single
question.

She could only repeat after the doctor, "Hôpital, hôpital," the
French term for hospital. She had no idea that it was Tuesday
morning in the summer of 1987, that Ronald Reagan was president, that we were living in America and that she had landed in
a Fifth Avenue institution called Mount Sinai that had no connection to God or Moses and was nowhere near a mountain. She
looked so lost and I knew—*I knew*—that she was mortified.

What the doctors failed to see was that while Mom's memory was gone, she still had the ability to feel.

After the doctor had left the room, after he had flashed a smug, self-satisfied smile my way as if to say, see, we told you so, we told you that she was a lost cause, it is no use, I would put my arm around her and hold her and try to comfort her, my poor child of a mother.

I knew I had to take action—I had to counter the effect of the corrupt Quiz Show. That is when I decided to administer a cognitive function test of my own. It wasn't too different from those long-ago exams she would give me, when she'd been so anxious for me to attend the lycée.

Edith, I would say, what is Gustave Flaubert's greatest novel? I gave her a small hint—*"Madame B. . . ."*

"Bo-va-ry," she replied shyly, mouthing every syllable.

"Who was the author of *A La Recherche Du Temps Perdu*? Marcel P. . . ."

"Proust," she said more confidently.

And Edith, what was Stendhal's greatest novel? At first, I didn't offer her clues; I wanted to see if she could answer it unprompted. For Mom, Stendhal was in a class of his own and as I was growing up, she was always quoting passages from his novels.

Edith, I repeated, Stendhal's most famous work? I saw her straining to remember, to remember. I realized I was being as heartless in my own way as the Quiz Show docs. Finally, I whispered, *"Le Rouge . . ."* (*"The Red . . ."*).

"Le Rouge et Le Noir," she exclaimed (*"The Red and the Black"*), and her voice was booming, and she seemed stronger and more animated than she had all morning, and certainly more cheerful than during her encounter with the doctor.

I saved my favorite question for last. Who is the author of *Goodbye, Columbus*—"Philip . . . ?"

"Phillipe Roth," she called out, making Roth sound French.

"And Philip Roth's aunt in *Goodbye, Columbus*—what did she love to do, Edith?" I pressed. "She loved to take out the . . ."

"The garbage," my mother answered immediately, and she was laughing and I was laughing as we remembered this favorite book of ours, and the section she had loved above all others, perhaps because she, too, was always taking out the garbage back on Sixty-Fifth Street. She would lug brown Key Food bags filled with trash down to the basement several times in the course of an evening, and I'd tease her.

Even then, she was able to laugh good-naturedly about her compulsion. Now I found she could still laugh, that merely by conjuring up Roth's fictional aunt, both the doctors and their fraudulent Quiz Show were at last forgotten.

When nothing I said could persuade Sinai to let her remain another day, I bundled Edith up and took her home.

We had hired an aide, Sondra Majors, an African American woman, tall and serious, with a quiet, soothing manner to her.

She was waiting for us at the door when our car pulled up.

"Edith," she said, taking my mother's hand in hers, "my name is Sondra, Son-dra."

I saw Mom flinch and her eyes widened. It was her own mother's name, Alexandra—*Alec-Sondra*—that she was hearing.

"Alexandra," Mom repeated, "Alexandra," and it was as if she had come alive again.

To her mind, my grandmother Alexandra of Alexandria had come back to take care of her: At last, she would have some relief from her agony. And in the weeks that followed, whenever Mom would get upset, only one person seemed to be able to calm her down, Sondra-Alexandra, returned from the dead to look after her daughter again.

Edith did make a comeback of sorts. Simply being back on Sixty-Fifth Street seemed to help. Little by little she walked in the street on her own, without the cane or walker the hospital had given her. But her gait was unsteady. She traipsed about like a drunken sailor, as if she were about to topple over. Sometimes César or I would come home in the middle of the day and find her lugging groceries from Key Food. César would watch her fearfully as she tripped across Twentieth Avenue, expecting her to fall down and send the bag filled with breadsticks and eggs flying to the sidewalk.

She had aides who were supposed to be doing this type of errand, but she tended to dispose of them fairly rapidly. To her mind they were all "des voleuses"—thieves, as she called them in the journal she was faithfully keeping. But there was nothing of value in the house, so what were they stealing? In truth, they were robbing her only of her privacy, her dignity, her sense of self.

The only aide she ever fully embraced was the quiet, mysterious Sondra-Alexandra. But she left us, too, one day, the way they all did—simply disappeared. She went on to another assignment, another sick old lady, I suppose, perhaps one with fewer emotional entanglements. My mother had no use whatsoever for the stream of women who followed. She would bark at them, grow impatient with them, order them to leave.

They drove her to distraction, these invaders of hearth and home. They made her anxious and restless; she felt as if she were jumping out of her skin. She poured out her anguish in the pages of my used reporter's notebook. Why did she need someone at her side? How could she keep coming up with chores for them to do?

That was the worst: She had to keep them busy. But she didn't have laundry for them to wash every day, or new groceries to buy every morning, she complained. What was there for them to do, these marauders who had taken over her home and wouldn't leave her in peace.

"These women the agency keeps sending will rob me of my last nickel," she raged in one diary entry. She veered between anger and sadness, between paranoid delusions and an odd clarity about what she was facing. Life was a daily effort to reckon with *la strocke.*

She was still preoccupied with my father, and in the midst of her own ordeal, she tried to figure out ways to reach out to him. She couldn't visit the nursing home anymore, at least not on her own—she depended on us to take her by car. Yet she constantly fretted about the shopping she needed to do for him, the grapes and oranges he would enjoy. The doctors may have decided her mind had lost its lucidity, but I found by reading her diary that she had a searing understanding of what had happened to him, and what would one day become of her.

She was perfectly aware how lost they both were. Though she was more than twenty years his junior, they were now sailing in the same perilous boat, she realized, their destinies as closely aligned as on that day in the spring of 1943 when they were married at the Gates of Heaven.

"Quelle vie??" she scribbled at one point. "What sort of life can I lead? It shall either be like this until I die, or else a nursing home like poor Leon. I have la strocke." She had a feeling she couldn't really depend on us, "les enfants"—the children—as she still called us.

César was consumed with his marriage and his wife and young daughters. Suzette was living somewhere in Europe.

And I was so busy, always working, always at my newspaper.

My new job at the *New York Post* was a nerve-racking affair. Editors were constantly sending me out to cover the latest gory murder. I'd run out the door to get to a crime scene, hoping against hope I'd persuade a neighbor to talk about the victim or, even better, offer me a picture we could put in the paper. It was a kind of street reporting I had never done, and which terrified me. Yet it was also a relief to lose myself chasing the crime du jour.

The *Post* newsroom was very sociable and convivial. Many of the reporters had worked together for years, but they were still embracing of me, a newcomer. I was unfamiliar with the gritty tabloid style but found I could ask some of the veteran journalists to help me. My favorite was a serious veteran reporter whom everyone called "Feiden." It was de rigueur to use last names in the newsroom, but he seemed to enjoy a hallowed status, the way star reporters do. Feiden—his first name was actually Douglas—was so good-natured that even in the midst of a deadline, when most of our colleagues were harried and on edge, I could turn to him for help and he'd leave his terminal, walk over to mine, and calmly review my story to make sure it was in the requisite punchy tabloid style.

There were times I didn't even have money for a cab to get to an assignment. Feiden would willingly take out a wad of bills and hand me ten dollars, twenty dollars, no questions asked. Occasionally, he and the other reporters would invite me to join them for drinks at the Lion's Head, a bar in the Village that enjoyed an almost cultlike status (though it seemed rather ordinary to me). I'd go for a while, pretend to sip a rum and coke, then race home to Mom in Brooklyn.

One day, Donny Sutherland, the amiable and manic news assistant, called out that there was an Edith on the line. When I picked up the phone, I was amazed to find it was my mother calling, wanting to know how I was doing.

She had found my number and then dialed it by herself—two distinct and, to my mind, wondrous achievements. We chatted about how her day and mine were going. I don't think she realized it had been months since she had made a phone call on her own.

I thought of the doctors' dire prophesies, their warnings that she would only deteriorate. That night, as I shared a taxi with Feiden to the Lion's Head, I confided to him what was going on with Mom and repeated a line I'd heard recently from a rabbi: "After the destruction of the Second Temple, only fools became prophets." He nodded, and I had a feeling he knew exactly what I meant.

When I came home later, I found Mom asleep in the little daybed I'd bought for her. Her skin was so smooth and silky, I noticed, a young girl's skin. She had always been so vain about her face—boasting that she didn't have a single wrinkle, *pas une seule ride*, a phenomenon she gleefully attributed to the fact that she'd never worn makeup. In the period leading up to her illness, she often seemed pale and wan, but now, after months of rest in effect, she looked prettier, healthier.

She seemed oddly content though I'd woken her up, and she called out to me "Loulou, Loulou," then fell asleep again.

My shift at the *Post* didn't begin till the afternoon, so we would have breakfast together, and I'd linger until the aide arrived. I had arranged for Mom to get Meals on Wheels—a Jewish nonprofit would deliver a ready-made kosher lunch and place it in her hand. After a few days, I noticed that she didn't eat it—not much of it anyway. Instead, she would hoard the tray—after a bite or two she'd put it away carefully in the refrigerator, not realizing that she didn't have to be so frugal, that another meal would arrive the next day.

"Je le garde pour Leon," she said when I confronted her; I am keeping it for Leon.

I had begun to feel restless again, and on weekend afternoons when I wasn't working, I would wander around the East Side. I was looking halfheartedly for an apartment. That old demon, the longing to run away, to flee, had returned to haunt me. I had also become seriously involved with "Feiden" and longed for a place where he and I could be alone.

Ours was a courtship on the fly, squeezed in between deadlines and assignments and my attempts to care for Mom. Feiden was a creature of New York, but a New York I didn't really know, that hadn't figured at all in my upbringing. He was an habitué of dim sum joints in Chinatown and pizza parlors in the South Bronx, and he was known to linger till one or two in the morning in various bars around town. He loved Little Italy because of a strangely affectionate bond he enjoyed with John Gotti, the formidable mobster and Mafia chieftain. Feiden had covered Gotti for the *Post* and was one of the first journalists to report on his notorious criminal exploits. He had dubbed him rather fancifully the "Dapper Don" in his stories and the moniker stuck. Feiden was fascinated with Gotti and proudly loved to show me all of his hangouts. On one of our early dates, he invited me to Taormina's, a restaurant Gotti frequented on Mulberry Street.

I didn't realize I was being taken for an audience with the Godfather.

When we entered, we spotted Gotti almost immediately. There he was, the legendary crime boss who looked like a movie star, enjoying dinner with his favored lieutenants. He was even more striking in real life than in the news photos; in his beautifully tailored silk suit, he made me think of Dad in his old salad days. Gotti's hair was perfectly groomed and his manner was oddly amiable for a thug and an outlaw.

I noticed that he was eyeing us—me—carefully. As we ordered dinner, a waiter materialized with two glasses of sambuca he told us were on the house. We looked up to see Gotti smiling and tipping his glass our way. I am not sure what I found more thrilling—being out with Feiden or getting a smile from the intensely charismatic, handsome Dapper Don.

I had the feeling I was being introduced to someone Feiden considered deeply important, someone whose blessing he sought.

Feiden and I were from the start inseparable. We worked closely together at the *Post* newsroom—though as a senior reporter he enjoyed all the prestige and confidence I lacked. After deadline, we'd go out either alone or with a group of other reporters for drinks and dinner. Come midnight, the first editions of the *Post* would reach the local newsstands and Feiden would run to get one and together we'd look for our stories and bylines. If one of us had the front page, or "the wood" as it was called in tabloid lingo, we'd

Loulou with Douglas Feiden; the two were married on New Year's Eve, 1996.

be especially jubilant and stay out even later, until he'd escort me back to Brooklyn.

There was my life with Mom and my life with Feiden, and somehow the two increasingly didn't seem compatible.

At home on Sixty-Fifth Street, I was getting into running arguments with my brother Isaac. He constantly pointed out faults with aspects of Mom's care, and I was worried that his critiques were a prelude to a greater, more sinister goal—he was going to find some way to force Mom into a nursing home, exactly as he had done with our father.

That was the American way—cold, ruthless, practical.

I rented a small studio on the East Side and continued to come home when I could—after work, on weekends. I persuaded myself I could keep a close enough watch to make it possible for Edith to stay at home—that I could supervise her aides, her meals, her medicines, her moods, even as I held on to my job at the *Post* and carried on a romance. I was no different than millions of self-absorbed, misguided American children, I suppose, who think they can have it all—that they can go on with their lives, hold on to their independence, flourish in their jobs and careers, even as their parents age and crumble and break apart with no one on hand to put them together again.

I may have been deluding myself; Mom wasn't.

One day, as I leafed through her diary, I found an entry that left me shaken.

She began, as was her wont, by noting that she'd taken her evening pills. Then she launched into one of her familiar tirades against the latest home health aide. She referred to her as "la bonne"—the maid—as if the woman Medicaid paid to care for her was one of the servants she'd hired and fired once upon a time on Malaka Nazli Street.

She had been to the nursing home that day to see Dad and

had come back agitated. My father hadn't wanted any of the fruits and cakes she had offered him, and that seemed to throw her into despair.

"Quelle affreuse vie," she scribbled—what a horrible life—"et bientôt ce sera la mienne"—and one day it will be mine. The entry was painstakingly neat and clear as if by some miracle of rehabilitation, she had finally recaptured some of the lost art of penmanship.

· 19 ·

The Woman Against the Wall

It was the bow everyone noticed—the enormous, shocking pink satin bow artfully positioned on the halo of dark hair. Some did a double take, pausing to look more closely at this *triste* elderly woman in the faded blue hospital gown who couldn't talk or move or even breathe on her own. Why on earth was she wearing such a festive accessory in a hospital room?

By the early 1990s, Mom had become a fixture at Mount Sinai—not in its rehab unit, but in its vast, hopeless wards. The doctors-prophets-fools had been vindicated, I suppose. Edith had deteriorated in ways that I couldn't seem to control, no matter how hard I tried. With the loss of her faculties, her body would also break down so that she kept getting sick, developing mysterious fevers and viruses and infections, and there was no option but to take her to the hospital. I would ride in ambulances with her and then ride back in ambulances.

I learned to find comfort in the sound of a siren.

Mom's own prophecy, carefully noted in that makeshift diary, had come true. She and Leon now led parallel lives, neighbors in an assortment of nursing homes and hospitals. There was nothing more they could do for each other and nothing more that she could do for him.

When César or I visited them, we'd put their chairs together side by side, but they still ignored each other.

Both now lived at the Jewish Home and Hospital, one of those large institutions that had proliferated in the second half of the twentieth century to meet the needs of an increasingly fractured America—an America where children moved away and parents and loved ones were left to fend for themselves.

Billions of government dollars were pumped into nursing homes like Jewish Home to provide the care families no longer did. No matter that in reality these institutions provided perfunctory care, and the people forced to live within them suffered at the hands of indifferent nurses, overworked aides, and arrogant administrators.

These institutions loved to brag about how large they were, as if size signified quality. They referred to themselves as five-hundred-bed nursing homes, never as "five-hundred-patient" or even "five-hundred-room" nursing homes because that would have meant the elderly were the priority, when it was in fact only about beds—filling beds, then refilling beds when they became empty, which happened constantly since people were always dying.

I was still waging the war that had begun in Sinai's rehab unit back in the summer of 1987. It was now more a form of guerrilla warfare, where I'd lob grenades into an immense black hole of a system that seemed impervious to my attacks. I fought on two fronts: in the wards of Mount Sinai and at Jewish Home, where

Edith was sent back whenever she was deemed "stable" enough to leave the hospital.

My fiercest attacks were leveled at the nursing home. My brother Isaac had chosen it—Jewish Home was located close to his Upper West Side apartment—and I had been unable to stop him. Besides, he seemed so sure of himself as he touted its Manhattan standards of efficiency and care.

These were the standards.

Every morning, my mother would be woken up at dawn, at six in the morning or earlier, forced out of bed, bathed, and dressed. Then she'd be made to sit in a wheelchair in the hallway until breakfast was served, hours later. Of course, she'd be asleep by then—most of the patients were—too tired to drink the watery, tasteless coffee they served.

Why, I demanded, did Mom have to get up at such an ungodly hour—why not let her sleep a bit more? Impossible, I was told—the night shift left in the morning, and part of their job was to pave the way for the morning shift. This meant that patients had to be roused out of their sleep and washed to lessen the workload for the next group of aides and nurses who trooped in.

That is when I grasped the fundamental truth behind nursing home life—it was driven entirely by staff convenience. There was a cruelty to it all—a viciousness—emblemized by the routine of waking fragile old people every morning at dawn for no purpose, simply to have them sit in a row in a long hallway.

Yet Jewish Home, perhaps more than other nursing homes, was masterful at maintaining a façade that hid the bitter reality. A large poster-size photo of Diane Keaton with an elderly resident hung prominently in the facility. "Diane Keaton loves it here," I was told, and I did indeed catch a glimpse of the actress one night, furtively coming to visit a patient who had also been a performer.

It was all so seductive, like the gleaming fish tanks in the visitors' area, and the make-believe coffee shop whose fare was nearly as dreary as what patients were forced to eat upstairs, and the library that was invariably empty. These gimmicks worked for a time on us, too.

Edith, enfeebled by several years of strokes and seizures, couldn't cope with the harshness of her new surroundings. She rapidly deteriorated and whether I came in the morning or at noon or in the evening, I would find her slumped in her wheelchair, fast asleep. She spoke to no one and barely ate or drank.

My mother had lost any semblance of an identity; she was simply a woman against a wall.

I was actually banned from coming too early. My visits were disruptive to the staff, I was told, and so most often I had to wait till lunch to see her. I'd tap Mom lightly on the arm to get her to wake up, then coax her to eat whatever lackluster offerings were on her tray—some sips of lukewarm soup, a couple of bites of the kosher TV dinner we'd insisted the nursing home, Jewish in name only, it often seemed, provide for her.

Edith was often made to sit near the nursing station, where she could presumably receive more supervision and attention. But I realized it didn't really matter. I would usually find the head nurse absorbed in her paperwork. Aides would walk in and out of the station and simply ignore her.

She wasn't even human to them, my exquisitely human mother.

She was only a patient in a wheelchair, exactly like all the other patients in wheelchairs.

I felt powerless to make any changes in her life. The nursing home tended to dismiss whatever complaints I made because I was "the daughter" and they expected daughters to be difficult and had learned to tune them out and to keep doing exactly what they were doing.

At least Dad in his chair in another hallway—the nursing home didn't put him together with Mom—would scream at the nurses and the aides, defy them, insult them, use choice Arabic curse words to make his displeasure and anger known. But Edith, the porcelain doll of Sakakini, raised from a tender age to be sweet and obedient, simply sat there slouched in her chair against the wall, unable to fight back.

Sometimes we simply grabbed her and fled. I became a master at demanding, and obtaining, day passes and afternoon passes and overnight passes from the Jewish Home apparatchiks. Together with Feiden, who now lived with me, we'd bundle Edith up in our rented car and take her somewhere—anywhere—far from the rancid air of West 106th Street. He would chant "Precious cargo, precious cargo" as he helped tuck Edith safely in the backseat like a little girl, like the child we didn't have.

We'd go for drives, picnics in the open air, to Fort Tryon Park and the Cloisters, to the Hudson River, locales she once loved and we thought would cheer her, but she was so quiet on these journeys. She'd watch silently from her window seat, a bit anxious, a bit on edge. I'd sit with her in the back and try to reassure her.

"Isn't Douglas an excellent driver?" I'd ask, and she would perk up and reply that yes, "*Dooglas* is an excellent driver."

One morning, we set out for Columbia University, which I hadn't visited since that year as an exchange student. The main entrance on Broadway was always closed to traffic, yet when we explained to the guard that Edith couldn't walk, he took the unprecedented gesture of opening the heavy metal gate and allowed us to drive into the campus.

Once inside, we put Edith in her wheelchair and pushed her around the quad. As we pointed out Butler Library—*la Bibliotheque*—and other grand buildings, she seemed happier and more

Loulou with Edith on a visit to César's house, Queens, late 1980s.

animated. We sat on the steps, near the statue *Alma Mater,* and parked her wheelchair. Students kept stopping to chat—they seemed to find Mom utterly adorable. She basked in the attention and answered them as best she could with a trace of her old verve, and I knew that she was brought back to those joyful days when I was a student, and she'd come visit me and bring me groceries and supplies, and life made sense.

It hadn't made sense in so long.

One Saturday, César invited us for lunch and we took Edith and drove to Queens. Once at my brother's house, we placed Mom in a chair at the end of the dining room table. She seemed so sad, and I tried to draw her out by talking about Cairo and L'École Cattaui. Even a mention of the school was usually enough to put her in a more cheerful frame of mind.

This time my little ploy had the opposite effect. Mom became intensely agitated. She sat up ramrod straight in her chair and her hands were trembling. Her voice rose as she began to talk about the key—the key to the pasha's library.

I was surprised; it had been a while since I'd heard about the key. "La clef, Madame Cattaui Pasha m'a donné la clef, elle l'a mis dans ma main"—the key, Madame Cattaui Pasha gave me the key, she put it in my hand, she said.

She spoke with pride and anger and a kind of wild, searing grief. As I looked at her sitting there, so fragile and careworn in that dining room in the middle of Queens, I understood her anguish, how far she felt from Sakakini and the grandeur of the pasha's wife and that time in her youth she had cherished above all others on this earth.

Out of the blue, we heard from the Brooklyn Public Library. They were inviting Edith to the annual Christmas party and seemed anxious for her to attend. The library's director, Larry Brandwein, was hosting the affair, and her former coworkers from the Catalog Department would all be there.

The morning of the party, I had an aide help me dress Edith in her nicest outfit. We wrapped her in a heavy wool coat and a scarf, then began the long journey from the Upper West Side to Grand Army Plaza. As Feiden drove, I kept telling her we were going back to the library—her library—but she seemed more confused than excited.

"Est-ce que tu te souviens?" I asked when we reached the vast traffic circle that she had once crossed intrepidly to get to her job. "Does this look familiar?" I pointed to the large building in the shape of an open book, once her home away from home. If she remembered, she didn't let on; she simply stared and I could tell she was nervous. But once inside the library's grand lobby with its familiar bronze doors and sky-high ceiling, she seemed to relax.

We made our way upstairs and then she was back—back in the cramped Catalog Department where she'd spent so many years.

Her old colleagues were milling about, drinking soda and munching on cakes and cookies. It had always been such a sociable group, cerebral as well as outgoing. I spotted Docteur Alexis, the tall, somber Ukrainian lawyer who had been her first supervisor, and was introduced to the soft-spoken Rabbi Sam Horowitz, the Hasid from Borough Park Mom had loved. Rabbi Horowitz had arrived on the scene a year or two after Edith had started at the library and they'd become fast friends: how she enjoyed consulting him on religious questions. Madame Mudit, her Latvian companion, was also there.

Then I spotted Dallas Shawkey, her former boss and nemesis, and my heart sank.

One after the other, her coworkers came over and embraced her and tried to engage her in conversation. She'd smile, but it was often a blank smile, as if she couldn't quite place any of them, not even Docteur Alexis, or Mudit, or Rabbi Horowitz, or Shawkey, the man who had driven her out of her beloved cubbyhole.

Though her friends welcomed her back effusively, I could tell they were unnerved by her appearance. The gaunt figure in the wheelchair had almost nothing in common with the elfin woman they had known and whom they'd loved for her wit and incandescent mind. I noticed Docteur Alexis staring our way; I had expected him to be the friendliest of them all. But he was oddly distant and reserved. I was puzzled—of all her coworkers, she had loved him the most; once upon a time, not a day went by that she didn't come home bubbling over with stories about Docteur Alexis. Was he revulsed or simply spooked?

Perhaps he was more shaken than the rest, I decided, and too honest to pretend nothing had changed.

After the last soda bottle was consumed, the last cookie eaten, and everyone had returned to their desks and their *Union Catalog* volumes, we had nothing to do but turn around and leave.

As we wheeled Mom out, we took the scenic route, through the library's ample circulation rooms with their shelves of books and honey-brown wooden cases containing the catalog cards she had once typed with such joy and fervor; I was praying that simply being in these familiar chambers would spark a memory.

Suddenly, we heard someone calling, "Edith, Edith," and a small man with blond hair came running over to us.

"It is me, Ed," he said, as he leaned over Mom and hugged her, "Ed Kozdrajski."

My mother frowned and then broke out into a broad smile—at last, someone she recognized amid the blur of unfamiliar faces.

"Ed—Ed Kozdrajski, is it really you?" she cried, voluble for the first time that day, and I had the sense that the shy, handsome young cataloger—the one that she'd always called "Le gentil petit Ed," who had been among the first to welcome her to Grand Army Plaza some twenty years earlier—had finally broken through the layers of neurological damage, so that she remembered.

We helped Mom into the car and began the drive to Manhattan and Jewish Home, where none of us had any desire to go.

I arrived one morning to find Edith in her usual spot against the wall. She seemed to be asleep as always except that she also looked extremely pale and when I tried to wake her, she didn't respond at all; her eyes were also out of focus, and I started to scream—she was either having a stroke, there, in front of us, or had already suffered one, and no one had even noticed.

The nurses and the aides had simply walked by her the entire morning, pushing their medicine carts, doling out pills and juice, without even realizing that one of their patients was critically ill.

I shuddered to think of the possibility that an aide had roused her out of bed and forcibly dressed her even when she was in the throes of a major seizure.

The charge nurse was in the station, still intent on her paperwork.

She didn't even look concerned when I confronted her. She'd no doubt have simply kept on doing her paperwork if I hadn't ordered her to get an ambulance at once, if I hadn't screamed at her at the top of my lungs.

At Sinai, the seizures continued for days and nights on end. Mom was placed in a special neurology intensive care unit. It became my home. I would sit there and watch the doctors walk in and out. Nothing the house staff was trying seemed to work. Couldn't one of these distinguished-looking physicians make the seizures stop? Couldn't they please make them stop? I was so desperate I took to grabbing doctors I didn't even know as they came in, and I begged them to please, please take a look at Mom.

It took days for her to stabilize and when she did, she was even weaker—more "compromised" as doctors said—than before. This latest blow left her unable to breathe or swallow. The hospital suggested a feeding tube, because it was now too dangerous to let Mom eat or drink. Then, later, I was told she needed a tracheotomy, because she couldn't breathe on her own either. Ultimately, she became dependent on a respirator.

As the days stretched into weeks, I took to coming early in the morning to the hospital to get her dressed: Sinai was more liberal—more human—about visiting hours than the nursing home, and I was free to come and go pretty much as I pleased and stay almost as late as I wanted.

One morning as I was helping the nurse get Mom ready, I decided to put a pink satin bow in her hair. Bows were in style, and street vendors on every corner were hawking them. I did it at first simply because I thought she looked so pretty wearing it. Then I realized that it helped her to stand out, and it became a powerful weapon in my arsenal, a way to get Mom the attention she needed.

Suddenly Edith was transformed; she was no longer merely another patient—she was now "the lady with the pink bow." The bow became a conversation piece as did she. Nurses seemed enchanted by her, and in the midst of their harried shifts, they gave her extra doses of TLC. Doctors came to check on her, and some actually stayed for a while, struck by her loveliness, her sweetness, which the bow brought into relief. That is when I realized that the disease we were fighting, the illness that had consumed her memory, her mind, her ability to walk and talk and even breathe, had left her with one crucial quality that no one could take away and would yet help her survive—her wondrous and bountiful charm.

*W*hy not take Edith home? As Mom lingered in the hospital and fell prey to countless mysterious infections, we were assigned a new social worker, an earnest young woman named Cathy who became my friend and confidant. I told her what had happened at Jewish Home—the countless assaults on my mother culminating in that last awful day in the nursing home hallway.

The thought of taking her back there was unbearable.

Besides, it wasn't even clear she *could* return. The horrible irony of nursing home life is that some patients become too sick to live in the average institution. Once they become dependent on a respirator to breathe, for example, they stop being desirable.

"Why not take her home?" Cathy said one morning.

The question hung over the dinner table at the restaurant where Feiden and I retreated after a long day at the hospital. He repeated Cathy's question, then added his own suggestion: "Don't let her return to the Jewish Home and Hospital—let us take her home."

He had recently bought a small duplex for us on the East Side. This meant Mom could have her own room upstairs while we would have our own private living area below her. Only a small spiral staircase would separate us, so we'd keep a watch on her at all times.

With his usual flair for words, he proposed transforming our apartment into "a hospital-free zone." I wasn't sure what that meant exactly but it sounded exciting, if sobering: We'd be re-creating a medical intensive care unit in our living room, stocking it with a hospital bed, IV poles, oxygen tanks, gauze bandages, blood pressure kits, saline solution, whatever Mom would need.

I worried at first about my siblings—would César and Isaac and Suzette go along with my plan or put up a fight?

I realized it didn't matter. I was driven by one goal—never to see Mom in a chair against a wall on West 106th Street again.

Cathy suggested an attorney who specialized in the elderly and knew how to fight entrenched American systems. Of course, my brother Isaac hired a lawyer of his own to stop me. In the past that would have been enough to make me cave, but now I braced for war. With Feiden and the social worker and the attorney on my side, I was ready to take Edith where she had always longed to be: at home with me.

I would finally try to rebuild a hearth—a fragile, antiseptic hearth perhaps, but still a place where we could be together again.

But how?

I started by going on a shopping spree. There was a Woolworth's on East Eighty-Sixth Street near my house that reminded me of the cherished dime store of my childhood. It even had bins with sale items in the basement. In a surfeit of optimism, I bought sheets and pillowcases and rugs and towels and a pink-and-yellow flowered bedspread. I planned Mom's homecoming by shopping with abandon and foraging through every bin, as she and I had loved to do.

Mom came home in early December, on the eve of Hanukah, the festival of lights, the holiday of miracles. Years earlier, in our house on Sixty-Sixth Street, instead of lighting a traditional menorah, she would take a set of juice glasses, fill them with water and oil, and insert a floating wick into each. Then, at night, we would light the wicks in the glasses, which we arranged in a semicircle. Edith would put her hands over her eyes and whisper a prayer as we took turns lighting them. She was always so thrilled when the flames lasted through the night. "Nes," she would declare in Arabic, "A miracle."

We would find ourselves staring and staring at the flames reflected in the water, as if they contained a divine message.

There were no floating wicks on the Upper East Side, but I did buy an electric menorah from Woolworth's. It was plastic with small orange lights. That was the last purchase I made prior to her homecoming, after the boxes of surgical gloves, the syringes, the medicines, the cans of liquid nutrition, the creams and salves I'd realized were necessary to rebuild a hearth.

· 20 ·

An Earthquake in Cairo

It was as if the real world didn't matter anymore—not to me and certainly not to Edith.

We hunkered down in the little duplex Feiden had purchased. He and I stayed in the basement bedroom while Mom's retinue of doctors, lawyers, nurses, aides, physical therapists, medical equipment salesmen, oxygen tank dispensers, and Medicaid fraud investigators roamed the living room and garden and small kitchen.

They would come and go, come and go.

Home felt like a way station these days. I would arrive and find four or five people standing around Edith, talking intensely, taking notes on a clipboard, inspecting her bed, her feeding tube, her.

Yet we were in our own way intensely lonely. Except for César who came every day and Isaac (less so), none of our visitors were friends or relatives—they were strangers who were paid to check on us, and they'd usually beat a hasty retreat after accomplishing their mission.

In spite of the stream of passersby, we were oddly isolated.

I didn't dare confide in friends that I had brought Edith to live with me. I was leading in effect a secret life—by day, a working journalist, a professional, and then at night caring for my mom. The few times I'd told people I knew, their reactions had unnerved me—they seemed to think it was all rather strange, bringing a sick mother home. It simply wasn't done, not in the solipsistic culture of Manhattan's East Side in the early 1990s, where everyone was intensely focused on their work and on themselves and not much else.

Even at temple on Saturdays, I failed to find the empathy I craved. On Sabbath mornings, I would bundle Mom up and together with an aide, we'd put her in her oversized wheelchair, her little oxygen tank tucked away in a back pouch, and call an ambulette to take us to a nearby synagogue. It took effort, but as my mother retreated more and more into her own world, getting her to a cozy synagogue to hear the chants and melodies she had once loved could only help her.

Or maybe not so cozy.

Typically, we went to the Fifth Avenue Synagogue on East Sixty-Second Street, which housed an Oriental-Jewish service for Levantine Jews and we also tried the Gates of Prayer, an elegant Reform temple nearby on East Seventy-Ninth Street. I felt uneasy and not especially welcome at either congregation. No one ever came over to ask how we were doing. No one came over to us, period. I suppose that we were a jarring sight for the well-heeled worshippers. Mom's wheelchair was a bulky affair, and it was hard to enter the sanctuary without attracting attention. I was intensely conscious of all the stares. Worst was the Gates of Prayer; I arrived there one morning with Mom in tow and was wheeling her toward the large sanctuary when one of the ushers raced over to say we had to sit in the back. He pointed to a dimly lit corner in the last row. I felt so humiliated it didn't even occur to me to ask him why we were being relegated to the back—surely it must be some mistake.

I had heard of black churches in Harlem and New Haven— some catering to the poorest families—that set the first rows aside for the disabled and wheelchair bound, placing them in front of all other congregants, and yet my own faith, whose rabbis preached a reverence for the frail and elderly from the pulpit, apparently hadn't thought of doing the same.

I suppose that it could have been different if I'd had friends in those congregations—people who knew me and would have made my mother and me feel at home—but in the years of my parents' illnesses, I had stopped being part of a community. I had become an itinerant worshipper. My childhood habit of going to temple Saturday mornings hadn't stopped—it is simply that I prayed at whatever hospital or nursing home Mom or Dad happened to be at on a particular week or month or year. I went to small anony- mous services attended by people who were ill and frail and pass- ing through, and whom I'd never see again.

I was haunted by memories of the Shield of Young David, where Mom and I had been embraced from the start and made to feel so welcome and loved: Wasn't that how a house of worship should be?

Ah, the synagogues of the Upper East Side—so upscale, such avatars of affluence and style, that prayer seemed almost beside the point.

They were our lone excursions outside the house, these forlorn trips on Saturday morning. The service at Fifth Avenue at least was tolerable—people were neither friendly nor unfriendly. But we never returned to the Gates of Prayer—I couldn't bear the thought of being relegated to the back.

The duplex on East Eighty-Ninth Street became our whole world.

There were those who were willing to join us, to help abate the isolation we felt, and those who wanted nothing to do with us.

I learned to depend on the *bon docteur,* that breed of physician who is both proficient and humane. Though house calls had long vanished from American medicine, several of the doctors who had taken care of Mom in her pink bow at Mount Sinai expressed a willingness to monitor her at home. They were specialists, with busy practices outside the hospital, yet I liked to think they came because they felt a genuine bond with my mother—though nearly mute and effectively motionless, she had captivated them somehow. They must have felt the intense life force within her, the will to hold on despite the blows that came raining on her one after the other. They were willing to apply their powers to keeping her if not well, then well enough to be home with me.

There was charming Dr. Stacy, our dapper Yale-educated neurologist, who visited us promptly every week. Dr. Kornbluth, a tall sober Orthodox Jewish gastroenterologist, was so meticulous when he examined Mom, as if she were a delicate flower that required all his skills. Dr. Dziedzic, our internist, had befriended us at Sinai and was our all-round, all-purpose medical sounding board, who took charge in a crisis. Her pulmonologist, Dr. Adler, came over to make sure her muddled lungs were clear and that she was getting plenty of oxygen. He spoke to her so tenderly I could swear she both heard and understood him.

She returned all their caring by a flicker of a smile, a gaze, or the lone word she could still say: "Okay."

"How are you, Edith?" Dr. Dziedzic would call out when he arrived. He was cheery and down-to-earth, devoted to his patients and to us. At first Mom would look at him bewildered, her wonderfully expressive eyes widening—who was that scary man? Then, reassured by his kindly manner, she would mouth, "Okay," her lips still slightly twisted from the aftereffects of all the strokes, the sound barely coming out.

I often turned to Dr. Diamond, our old neurologist, for advice. Mom was sleeping a great deal and she was barely speaking, I confided. He smiled, reassuring in a way almost no one else was anymore. There were people who could convey more with one or two words than most of us could express with the entire English vocabulary at our disposal, he told me. That is how I had to think of Edith—not by focusing on her limitations but on what she managed to convey in spite of the limitations that God and destiny had inflicted.

Abandoned by our traditional friends, we made brand-new friends. Mitou, a tall pleasant Indian who worked at the drugstore around the corner, came over every day to help get Mom out of bed and into her wheelchair, and then returned in the evening to help us get her back into bed. Aldo, who delivered the canisters of oxygen, always lingered a bit, captivated by the sweet, silent figure in the bed. Even the occasional inspector from Medicaid, the health insurance program that was paying for Mom's elaborate care, would arrive prepared to be skeptical but left persuaded that the government was getting its money's worth.

Then there was the army of aides the agency sent over to look after Edith. These were mostly impoverished immigrants, typically from Africa or the Caribbean, who worked for a pittance—$5 or $5.50 an hour—on shifts that were typically twelve hours long, so that they were with us from morning to night, or night to morning. They had grueling commutes—from Crown Heights or Coney Island or Bedford-Stuyvesant or the far reaches of the Bronx— and those who reported for duty in the morning got up at dawn to be with us. In an economy that was flourishing, these women— and they were all women—were the only ones willing to take care of the elderly and the sick. They worked at the jobs no one else in

New York wanted, jobs that came with almost no benefits, that hadn't a shred of security.

They'd settle in the corner of the sofa across from Mom and seemed grateful for small comforts. They loved the old portable black-and-white TV set they could turn on between duties or while sipping the occasional bottle of soda or iced tea I tried to keep stocked in the refrigerator. Most led hardscrabble lives, raising children with the little they were earning, and often no husband to help them. Since they were used to jobs that lasted only a couple of days, they appreciated the steady work on East Eighty-Ninth Street. Mom's long chronic illness meant they were assured of a stable income for weeks, maybe even months.

I found them comforting to have around the house. It wasn't only the physical help they provided—the fact that they could keep Edith comfortable and clean. It was the knowledge they were watching over my mom when she could do nothing for herself anymore. Some became my friends and confidantes and threw themselves into helping me rebuild the hearth.

Like Clara Sylvestre, a young woman from French Guyana whom we'd known back in Brooklyn. Because she was fluent in French, she could connect in a special way—my mother had never stopped favoring the language of her youth. Clara would keep up a steady banter, talking about herself, her other patients, or my father whom she also looked after. She would gently try to remind Edith about Leon.

Had Mom forgotten him? She never seemed to react anymore when Clara mentioned Dad, when she said his name.

Margaret Mukoro, tall and striking, brought her boundless energy and verve into our increasingly lethargic household. A native of Ghana, Margaret had never known her own mother, and it was as if she were pouring all of her pent-up love on Mom. She worked the night shift. Every evening, at 8:00 P.M. on the dot,

she'd arrive and get to work. "Hello, Edith," she'd say in her booming voice. She was a whirling dervish of activity, changing Mom's
gown, pulling her up on the bed, trying to keep her awake by clat-
tering about the house.

Sometimes, my mother would look visibly distressed at all the
effort required of her. Margaret would say, "No more, Edith?,"
and we would see Mom painfully mouthing the words *no more*.
We knew when we were pushing her to the limit and we teased
her. "No more war," we'd chant. It was a nonsense rhyme, nothing
more than that, but it seemed to amuse her and make her smile as
she tried gamely to answer, "No more war."

How are you doing, I'd ask my lovely eloquent mother who had
worked for a pasha in Cairo and a public library in Brooklyn.

"Okay," she would reply.

Silence was a luxury, privacy nonexistent. Our house was constantly filled. People talk of the sacrifices to be made when caring
for a loved one but there weren't any—not really—except for the
difficulty of finding a quiet corner. There were times when I felt
besieged, and once I sought counsel from a therapist with a practice close by in the East Nineties. The psychiatrist seemed amiable
enough at first, but when I described to him my life at home, the
fact that I'd taken my mother out of a nursing home to live with
me, he began to yell. Why was I doing this, he wanted to know. "Is
she even alive?" he thundered.

I left in a daze. His attitude was extreme but by no means
unique. Our culture was obsessed with the notion of quality of
life, and it had become a euphemism for letting ailing loved ones
die without a qualm. Typical was the advice I heard from her old
social worker at Jewish Home. My mother "was holding on because of you," she said when she learned I was taking Edith out of
the nursing home for good. "You have to let her go," she told me.
There was to my mind a terrifying societal tilt toward mercy kill-

ing, and I knew I had to fight against it and all of its proponents. I had to fight as tenaciously as my old heroine, Emma Peel, had battled the bad guys. I had to be my mother's Avenger.

We favored mercy living on East Eighty-Ninth Street—the gentle touch of Clara, the sweet words of Margaret. They came from cultures where people still looked after their loved ones, and most had worked in institutions so they had no illusions about the level of care parceled out at the typical American nursing home. These women became my fiercest allies; they were with me on the front lines helping Edith survive, and I came to value their counsel far more than that of my peers or the so-called professionals.

No more war, I'd tell Edith, no more war, and I would see her attempt that crooked, intensely endearing half smile of hers and felt immediately heartened about the path I had chosen for us both.

I'd go to work at times with a sense of foreboding and race home in the afternoon, always fearing a new calamity. But there were weeks, even months, when Mom remained miraculously stable. I was even able to take holidays with Feiden—small getaways out of New York. César would relieve me and move in downstairs to supervise the nurses.

We flew to Montreal one week when Mom seemed to be doing especially well. It seemed appropriate somehow to choose Edith's favorite destination for our vacation, the place where she and I had once enjoyed a respite from a world that had become impossibly bleak.

Walking along Saint Catherine, I found myself searching for the hotel where the two of us had stayed in the summer of 1973. I couldn't remember its name or its exact location or even what it looked like, but still I searched. Somehow, merely being in this

French city had restored hope in both our hearts. "Live and let die," Edith had whispered to me that night at the movie theater on Saint Catherine, because she had the kind of mind that could find a powerful idea even in a James Bond film. We had prayed for a miracle at each one of Montreal's chapels and cathedrals and synagogues. I suppose that God heard us, and I was saved.

But what about Edith—she of the fasts and incense burnings and incantations? Didn't she also deserve a miracle?

I could almost trace her downfall to that summer, when she despaired of ever seeing me well again. My illness—that taboo subject she and I only tackled briefly, elliptically, in hurried, self-conscious conversations—had so consumed her that she was spent. She stopped eating, caring, looking after herself. She resorted to Alexandra's diet of Turkish coffee and sesame biscuits. Even as I went on a tentative road to recovery, her decline started—the decline that had led her to this state, lying in a bed in the upstairs of my duplex, unable to eat or drink or speak.

I wondered if the old canes and crutches were still there, at the shrine on the mountain, and if ailing pilgrims had continued to go to Saint Joseph's and climb steps and light votive candles and pray.

Did they still believe in miracles? Did I?

One October night when we were all together, Feiden and Mom and I, the aide flicked on the TV as was her habit. The news immediately caught our attention: There had been a major earthquake in Cairo; hundreds of people had been killed and many more were injured, and thousands of buildings all over the city had collapsed into heaps of rubble.

My mother's beloved city was in ruins. The death toll was mounting, buildings were crumbling everywhere, people were crying in streets that were no longer recognizable, that were overwhelmed with the dead and dying.

Our streets, our city.

When I looked at Edith, I saw to my surprise that she was sitting almost at attention. Her eyes were wide open, and she looked more alert and engaged than I had seen her in months—even years. The shocking images had brought her out of her stroke-imposed languor, and she seemed to be devouring the newscaster's every word and watching the television screen intently. While the doctors had spent years telling us how compromised her mind was, I realized that she had grasped the depths of the unfolding horror.

It was 1992; we had left Egypt thirty years earlier.

Yet it wasn't some faraway event in a distant land. The earthquake had struck our home.

That night, my mother was very agitated. She stayed awake trembling and crying and while I tried to wrap the quilt around her, she was still shaking, as if the ground that had opened up and swallowed her beloved Egypt was engulfing her as well.

We kept the TV on and each time the announcer would mention Cairo, I saw her flinch. I held her hand to calm her. I had the sense she was mourning the city of her youth—now as ruined and ravaged as she. Although the rest of New York had lost its meaning and faded in importance, Cairo remained deep inside of her—vibrant and alive and primordial.

BOOK FIVE

The Book
of
Lamentations

CAIRO, GENEVA,

JERUSALEM, AND BROOKLYN:

2009–2010

· 21 ·

The Verse of Consolation

Years after the Shield of Young David had ceased to exist, and I no longer wore jaunty hats and blazers with gold crests, and Edith wasn't at my side anymore, I would find myself looking back on that period of my life in the women's section wistfully and with tremendous longing. I would examine my past as a little girl seated next to my mother, peering intently through the diamond- and clover-shaped holes of a small and ornate wooden fence.

We were all dispersed by then, members of a thousand different congregations or in some cases, no congregation at all.

Occasionally, fragments of news would reach me.

David, the rabbi's pale young son, was diagnosed with leukemia and died. Marlene, my mother's gentle favorite, had visited him at the hospital toward the end. Gladys, who had watched over me from her perch in the kitchen, finally gave birth—quite miraculously—to a baby girl, years after she'd prayed and given up hope and continued praying.

By then, she and her husband, Saul, had adopted a child, a blond, blue-eyed boy the community suspected wasn't Jewish but he was. They named their baby Mark and doted on him, and then, a couple of years later, they adopted another infant, a girl. When Gladys learned that she was pregnant, her first thought was that her cries all these years in the women's section had finally been heard.

What became of Gladys and the other women behind the divider? What happened to the men beyond it?

That is what I found myself wondering after Mom died, a couple of years after I brought her home. She had gone into respiratory failure one night while Feiden and I were out having dinner.

Suzette, who had missed Dad's funeral, flew home the next day from London, where she had moved after California. At the cemetery where Edith was buried in a roadside grave next to Leon, my sister noticed there was a school nearby. Mom would like that, she said: She had always wanted to be around schoolchildren.

After Edith's passing, life seemed to lose its moorings. I bought the Brooklyn apartment she had loved so much—the house on Sixty-Fifth Street where she had longed to return after her illness—and kept it with her exact furnishings, yet no tenants. I couldn't bear to live there, and I couldn't bear to have anyone else live there either. I would go from time to time and remember her and the hearth she had so desperately wanted me to rebuild.

That is when I began increasingly to think of the Shield of Young David, and its women's section.

I would imagine myself sampling Gladys's tuna fish again, or watching her younger sister Fortuna getting all excited about the latest Elvis movie, or I'd find myself musing about little Maggie Cohen, clutching her father's hand.

And Celia, what about Celia? I was haunted by the memory of my friend's wedding—that long-ago night when I first realized I was sick. I had felt so tired I hadn't even danced, and Mom became worried. The next day we had gone to the hospital where the doctors ultimately diagnosed me with cancer. I'd wonder about Rabbi Ruben in Jerusalem: Did he still give stirring speeches? Were his daughters continuing to rail against Darwin?

And what about the Messiah? The Messiah that each and every one of us behind the divider had believed was around the corner. . . .

He still hadn't come. And somewhere along the line I had stopped waiting and stopped hoping and maybe even stopped believing.

In my new world, no one thought much about the Messiah, no one concerned themselves with his imminent arrival. Perhaps I didn't need a savior anymore: I was by any measure doing so much better than in my childhood, when my refugee family had barely coped. I had broken free of the women's section and lived far away from the strictures and petty rules of the Shield of Young David.

I had gone to work, not as a secret agent or an international woman of intrigue perhaps, but still very much as I'd wanted to do years back at the Shield of Young David, when I'd found the men's world so sober and serious, and the women's section so frivolous and unsubstantial by comparison.

I became a journalist, an investigative reporter, and I even turned myself into an avenger of sorts, always on the lookout for "extraordinary crimes against the people and the state," exactly like my childhood idol Emma Peel.

I searched for the Nazi war criminal, Dr. Josef Mengele, and my reporting on a group of his victims, the young twins who had been subjected to his medical experiments during the war, helped spur an international manhunt. Multiple countries, including the United States and Germany, launched a concerted effort to find the monstrous "Angel of Death" of Auschwitz. It was too late, of course—by the mid-1980s when Western governments at last began pursuing Mengele in earnest, he was already dead, having been able to remain in hiding all these many years with the help of his supportive Bavarian family.

Later, when I joined the *Wall Street Journal*, I turned my attention to seeking out corrupt hospital and nursing home executives who had, in their own way, inflicted untold suffering on the frail and the elderly and the vulnerable. It was all because of my experiences with Edith and Leon, of course. Every exposé I produced of an institution hurting its patients, every investigation into some failing in America's health system was my way of avenging my parents.

And when Rabbi Kassin was arrested in the summer of 2009— alleged along with other rabbis of engaging in money laundering— I was the journalist assigned by the *Journal* to do a story about him and my old community. In the years since I'd left, Saul Kassin had become a revered figure, known for his good works and charitable instincts, and he was crowned chief rabbi. It was the highest honor imaginable—a status some said should have gone to my beloved Hebrew school teacher, Rabbi Baruch, but he was passed over. When I heard of Rabbi Kassin's elevated status, I felt the same sense of bewilderment I had as a child when he'd rebuked me for asking why the Messiah couldn't be a woman.

I traveled to Deal, the Jersey shore town where Syrian and Egyptian Jews now summered. Many of my old friends and neighbors had prospered, and this elegant seaside community of large Victorian mansions was proof of their achievements in America, even as the ubiquitous kosher pizzerias and bakeries also showed me they'd stayed true to the ideals of our childhood. They'd even built a new synagogue called the Shield of David—Magen David of West Deal, a beautiful, imposing structure with a light and airy women's section.

But along with the wealth and beauty, I found suspicion bordering on paranoia. I had been received with open arms the prior year, embraced as the prodigal daughter for my book that was as much an ode to the community as it was a memoir of my father. Yet now that I was coming as a journalist, I was met with implacable hostility. "It is not personal," some said as they escorted me to the door, but it always felt personal. Nothing could persuade them that I planned to do a fair and loving story on this world of mine. And while I admired the opulence, how so many had attained the American Dream while shielding themselves from so much that was American, a piece of me mourned the simple ways of my community of old. And I was saddened to hear a year or so later that Rabbi Kassin had plead guilty to operating an illegal money transfer business. He was by then nearing ninety and had to agree to turn over several hundred thousand dollars to the government.

In my postdivider life, while I didn't use judo moves or karate chops, I was always on the attack, as Mrs. Peel had been. I had broken free of the barrier. I had fled the women's section and joined the men. I now sat side by side with them in newsrooms and corporate offices and boardrooms.

I was part of an entire generation that had escaped dividers. Even women who had never stepped foot inside an Orthodox shul—women who weren't even Jewish—had felt that same burning need as I did to demolish barriers. We had, together, reached the men's section and exultantly sat down with them to find . . .

To find what exactly?

That was the question, and I was no longer as sure of the answer as I had been in my arrogant years.

Now, I longed to know what happened to those who'd chosen the other path.

One by one, I learned, the girls who had been with me behind the divider had settled down, most at a very young age.

Gladys's sister Fortuna, well over her crush on Elvis, became engaged shortly after high school, at the age of nineteen. She had a family, including two daughters who became lawyers. She stayed close with Gladys to the end: My cherished protector from the women's section had died a few years earlier, I learned, as had her husband. Sweet Marlene married even younger, at eighteen, then surprised us all when she and her Israeli husband, Avi, helped found Bonjour, an international blue jeans empire. Marlene emerged as a woman of considerable influence and means, a pillar of her increasingly affluent community. Outgoing and chic, Marlene and her husband became walking billboards for the fashions their company marketed across the globe. She also assumed the role of matriarch, strong and purposeful. Her friends, siblings, and children turned to her exactly as we had as children behind the divider.

Maurice, the object of my girlhood adoration, became a businessman and never left the community. He remained close to Marlene and like her, he was devout and faithful to its values. Pursued by any number of women, he settled for a lovely blonde, and together they had a loving family. I could never probe: The few times I encountered him over the years, I felt as shy as when I was a little girl, enamored of a striking older boy.

Joseph Hannon, Maurice's best friend, seatmate, and confidant in the men's section, joined the Jewish Defense League, a radical organization that advocated violence as necessary to combat anti-Semitism as well as to counter the historical image of Jews as victims. Joseph became a devotee of JDL's founder, Rabbi Meir Kahane, and was said to roam the streets of Brooklyn carrying a lead pipe inside his jacket.

He eventually moved to Israel and became a religious scholar, a calling possibly more suited to his quiet, thoughtful nature. Now known as Rabbi Yosef, he roamed the streets of Jerusalem with his long beard and favored prayer books over lead pipes.

Celia, my untamable Moroccan friend, was also eighteen when she became engaged. We didn't know the boy, who came from outside the community. He had been raised by ultraorthodox Jews from Eastern Europe, and we weren't sure what to make of him. We had been taught at an early age to marry only "our own"— other Jews from the Levant.

At her wedding, on a bitter cold day in February 1973, Celia appeared in long sleeves, her face covered by her veil, and walked around her husband seven times, in keeping with his religious tradition (though not really our tradition). After some years, my friend—now the mother of four children—divorced her husband, an exceedingly rare occurrence in her world. She never spoke about the marriage or what had gone wrong. But she bravely reinvented herself as a professional woman. She ditched the hair coverings and long skirts of her former world and went to work. She held jobs at several airlines and began to travel widely; she even went back to visit her hometown of Tétouan, Morocco, which she had last seen as a little girl in the 1960s. When I found her, she had acquired a new identity; she was working as a massage therapist and looked striking in white slacks and sandals. We agreed she had come a long way from her strict House of Jacob upbringing.

Madame Marie, her mother, left America for good. She and her husband moved to Israel and, in late middle age, began a new life close to their youngest son. They settled in Jerusalem near

Moshe—Moses—who unlike his namesake had managed to reach the Promised Land.

That shy, stammering child—who had loved to wander into the women's section and longed to be included in our games—was now a towering figure in Israel's religious circles, a distinguished and prominent Hasidic rabbi. The author of numerous religious texts, Moshe was so revered that he taught other rabbis. He also outgrew his awkward childhood and found true love. He married a pleasant, thoughtful woman and together they had eleven children.

He and his family lived in a religious complex outside of Jerusalem, a community where even grocery stores were segregated by sex and women could only shop for supplies at certain times of the day, when there were no men around. His mother, always a voice of moderation, became more religious and began to cover her hair with a headscarf. She had never done that in all those years seated next to Mom and me in the women's section.

The Cohen sisters sold the house on Twentieth Avenue where I had whiled away so many Saturday afternoons and moved near Ocean Parkway. Rebecca Cohen was the first of the five sisters to find a husband. It wasn't easy, of course, because of the two disabled siblings, Leah and Maggie In a community that thrived on gossip, it was whispered that the Cohen girls were cursed with bad genes, and I think that my friends themselves at times wondered whether anyone would marry them.

Rebecca broke the curse. In her early twenties, she became engaged to an amiable Jew from Egypt named Eric Choueka, who fell in love with her and was willing to take a chance. She threw herself a lavish wedding and gleefully proclaimed herself "Rebecca Choueka." Strengthened by her husband's love, she proceeded to have several children, each one healthier and more promising than the next.

Her older sister, Gracie—who now called herself "Grace"—

followed suit some years later. She was a professional woman, a teacher in the New York City school system, when she went outside the community and wed a man who was Jewish but not Syrian. The Cohen sister I remembered most vividly and missed the most was Maggie. The sweet Down syndrome child, whose dearest companion was her father, had remained close to Abraham Cohen until one day while crossing the street he was hit by a motorcyclist and injured. In a surfeit of bad luck, gentle Mr. Cohen was also diagnosed with Parkinson's and could no longer take care of his cherished and most helpless daughter, the child who had loved him unconditionally. The sisters rallied and looked after their dad until the day he died.

Leah, the oldest, went into a group home with other young adults who suffered from disabilities. Later, I heard that Maggie had also been placed in a communal residence. But the Cohen sisters led fiercely private lives, and I never had news of any of them again.

Marlene's sister, Diana, my friend who had sat by my side every Saturday as we advanced step by step into the men's section, left the community and became an accomplished journalist and foreign correspondent. She married, divorced, and married again; and I sometimes wondered if she ever missed the world her sister emblemized.

Laurie, the princess of West 116th Street, who hadn't been with me behind the divider, but had prayed very happily behind her own partition in Long Island's Orthodox enclaves, decided to move to New York, to a grand apartment on Fifth Avenue. Now a mother of six and a grandmother of two infants, she welcomed me into her family in the same way she had so many years back when she'd invited me for weekends to North Woodmere and Hewlett Harbor. One night, as we were celebrating Passover together, Laurie suddenly stood up and began to dance. She danced around the long holiday table and she danced through the house and her four daughters danced behind her. They were all laughing and singing exuberantly and I had no idea why, but this demonstration of sheer *joie* reminded me of why I'd been drawn to Laurie so many years back when there was no joy within me and she had managed, miraculously, to restore a bit of it.

After the Shield of Young David, Rabbi Ruben found another calling: He worked as a stockbroker. He proved to be a gifted and talented investor, but then he grew disillusioned with Wall Street and the world of finance. He and his wife moved to an ultrareligious corner of Jerusalem, and several of his daughters settled there as well and raised large families of their own. My childhood rabbi reinvented himself yet again as a Hasidic scholar and Cabbalist. He traveled constantly to Safed, the ancient city of mystics.

For years and years, I had no news of my old nemesis, Mrs. Menachem. Yet she haunted me.

I could still hear her shouting, "You are a silly, silly little girl who's trying to change the world."

I would find myself increasingly wondering if she'd had a point.

Dalida, my mother's favorite singer, the former Miss Egypt who became a star at the Olympia and whose songs so vividly brought back the life left behind, took her own life one day in Paris with an overdose of pills. "Life has become unbearable—forgive me," she had scribbled in one final note.

The Congregation of Love and Friendship, my father's favorite synagogue, abandoned its modest home on Sixty-Sixth Street. It moved to grander digs in the heart of Ocean Parkway, and then, as Egyptian Jews prospered, moved again to an even more splendid sanctuary across the boulevard. The new temple of marble and gilt was so different from the simple structure Dad had loved. But the congregation did keep its original name, Ahaba ve Ahava, which conjured up both the old synagogue in Cairo as well as the little shul in Bensonhurst that had gathered everyone in from the cold.

The Synagogue Without a Name disappeared without a trace. The Reform temple I had never dared to enter was destroyed in the 1980s in a mysterious fire started by an arsonist. Years later, I met a woman who told me she had worshipped there as a child and yes, it was true: The congregation never had a women's section, exactly as we had suspected. But it did have a name—the Mapleton Park Hebrew Institute.

Only the Big Shul on Sixty-Seventh Street, next to the Shield of Young David, survived. Even as its congregation aged and dwindled, it limped along, holding occasional Sabbath or holiday services in its magnificent sanctuary. But then it, too, was forced to make a new life for itself.

It became a mortuary, where community members went to honor their dead. Whenever a Syrian or Egyptian Jew passed away, whether they were in a mansion on Ocean Parkway or in a medical center in Manhattan, they were transported back to Sixty-Seventh Street to prepare them for their reunion with God. The community had no use for traditional funeral homes, so popular in America, where death had become a business and burials were brisk and fast and impersonal. Not at the Big Shul.

A tall elderly gentleman who always wore dark sunglasses, even indoors, and whose name I never knew, stayed at their side, ministered to them, and recited the Psalms because the dead could not be left alone even for a minute. He washed them thoroughly and then dressed them in fine white gowns so they would be clean and pure on their heavenly journey.

For this tremendous *mitzvah*—good deed—the man with the black glasses was paid only a pittance, barely enough to support his family. He was also deeply feared. To his own eternal sadness, children ran away from him; and adults, even well-meaning adults, adults who should have known better, adults who did know better, shunned him as if he himself were the Angel of Death and rarely

invited him to their happy occasions, the weddings and bar mitzvahs he so longed to attend.

He was in fact a holy man, the dead's most tender companion—
the only friend they had as they marched off to meet their Maker.

Come morning, the man with the dark glasses left, and the
mourners arrived to take his place. They found their seats in the
sanctuary and lingered and prayed, and you could see them occasionally staring up at the ceiling, at that delirious light blue
painted sky above the altar, wondering whether their loved one
had found their way to it safely.

As I tracked down, one by one, these friends from my arrogant years, what did I learn? Marlene offered the most
powerful, and perhaps the most searing, lesson: When I saw her, I
realized that those of us who had sought to leave the women's section had paid a price far beyond our reckoning.

I found her in a stately home in Ocean Parkway, at the heart of
the community she had chosen never to abandon. Now a mother
of five and a grandmother of "more than" twenty—she was too superstitious to say exactly how many grandchildren she had—she
spoke with her old fervor about the world I had left and the world
I had embraced.

In my absence, "the Community," as she called it, had grown
and flourished.

Its members lived in elegant private homes and apartments all
around the wide tree-lined boulevard that had filled my mother
with such longing. Mom would always say so wistfully, "Oh, une
maison sur Ocean Parkway"—A home on Ocean Parkway—as if it
were the ultimate ideal.

Families stuck together here, and children lived near their
loved ones even when they were grown: that was the rule. Marlene

pointed to her own married daughters—lovely brunette versions of her—who had settled all around her, in touch with her constantly. Her grandchildren were her love, including a few who were named "Marlene" because that was the Syrian tradition: to honor your mother and father by naming your children after them. One beautiful little girl with fair eyes was nicknamed "MarleneBlue."

Above all, the Community took care of its own, my friend reminded me. If someone was sick and infirm, there were armies of volunteers rushing to visit them and comfort them and bring them soup. A bride in need of a trousseau could count on getting the fine clothes and gowns she needed lest she be embarrassed on her wedding day. A young man about to be engaged would be helped in purchasing a ring for his intended.

It was exactly as the Jews had functioned back in old Cairo and in long-ago Aleppo, as it had in the world of the pasha and his wife, when philanthropy was personal as well as communal and didn't depend on welfare or bureaucracies or the United Jewish Appeal.

And that outside world I had found so seductive?

It was a wasteland, a lost and hopeless place, my friend clearly believed. While she loved America, it was such a lonely country—so many American families were broken, fractured beyond repair. Children lived hundreds, thousands of miles from their fathers and mothers. Grandchildren hardly ever saw their grandparents. Families came together once or twice a year—Thanksgiving, Christmas—on what had become requisite, almost forced reunions.

But here on Ocean Parkway, where familial bonds still mattered above all, "We have Thanksgiving every week," she said.

It was the siren song Marlene had sung for years—every time I had run into her—a melody that filled me with yearning and where the lyrics consisted of only two words, come back come back come back come back.

had visited Israel many times, but I had never thought of finding Rabbi Ruben. For years, my life had been so removed from anyone who figured in my past. But suddenly I felt a deep need to see my childhood rabbi again. Invited to attend a professional conference in Jerusalem, I snuck out and traveled to his house in a religious enclave known as Ramot, the kind of area where families have large numbers of children and men are only seen in black coats and women always keep their hair and arms and legs covered and strangers are looked upon with suspicion.

I remembered my rabbi as vibrant and modern—a man of his times in the way he spoke and dressed. I almost didn't recognize the imposing figure with the flowing white beard who greeted me at the door. With him was his daughter Debbie, whom I'd known as a quiet little girl with fine blond hair. She, too, had been with me behind the divider, and like her older sister Miriam, she had shared her favorite books, including *The Kiddush Cup That Cried*. Debbie now wore a headscarf that hid nearly every strand of her hair, and she clearly lived a traditional, Hasidic lifestyle.

Once inside the apartment, I noticed the portrait of David, the rabbi's lost son, prominently on display in the dining room.

It was in the place of honor, where he could always see him. Nearby were pictures of Mrs. Ruben, the rebbetzin: She, too, had died. Although his daughters and grandchildren were all around, I had the sense that my rabbi was terribly lonely.

I wanted to know, how was Mrs. Menachem? I had heard she was living in Israel, though her children were in America.

She was gravely ill with cancer, the rabbi told me quietly. We could only pray for her. He didn't sound hopeful.

I asked father and daughter if they remembered a book I'd once borrowed from their family library—a work attacking Charles

Darwin. The mood in the room, subdued and wary since I'd arrived, suddenly turned jovial. Oh, you mean by Rabbi Avigdor Miller? You mean *Rejoice, O Youth?* both said at once. They were clearly delighted by my inquiry. Debbie made a phone call, and within minutes, one of her daughters raced over with a copy of *Rejoice, O Youth* in hand. There it was, the book I'd last seen as a little girl, with page after page challenging the theory of evolution and demanding to know where was the missing link.

I had joined Darwin's world in those intervening years. I lived among people who'd never even dream of questioning his theory of evolution, who thought it was heresy—madness—to do so, and despised the so-called creationists. Yet here I was with important figures from my childhood who still clung to notions I had discarded one by one, that I had come to regard as hopelessly quaint. The Rubens were more fervent than ever in their beliefs, and I am sure they were contemptuous of Darwin and so much else that my world held as iconic.

It had been more than forty years since I had flipped through the pages of Rabbi Miller's primer. I had since come across hundreds of articles, often on the front page of major newspapers, announcing some great archaeological find, the discovery of some skeletal remains that were heralded as definitive proof of Darwin's theory, and indeed said to be the "missing link." Somehow, they never were at the end—there was always a disclaimer. And that is when I'd find myself thinking of that long-ago book and wondering—*Where is the missing link?* That is when I'd realize that a small piece of me was still secretly, privately, quietly cheering Avigdor Miller and his legion of followers on the other side of the divider.

As I rose to leave, Rabbi Ruben motioned to me to wait: There was one more serious matter we needed to discuss. He pointed to my book, the memoir of my father I had sent him as a gift. He had only read the first several pages, but what he'd gleaned so far had disturbed him profoundly. Didn't I know that the most important commandment in the Holy Bible was to honor your father and mother? He suggested that I had shamed my dad—he accused me of dishonoring Leon's memory.

The evening suddenly took on a nightmarish cast. Surely you have to keep reading, I said, distraught, and you won't think this way as you read on.

But he wasn't listening, and I realized his mind was made up. I was, to him, the errant child, the silly, silly little girl of Mrs. Menachem's invectives who had wandered out of the women's section into a forbidden world and become so lost in it that I'd broken one of God's most sacred and essential laws.

I felt more triste than angry. I had journeyed thousands of miles, compelled to find a man who had made an indelible impression on my childhood, someone Edith had genuinely loved and respected, and whose lively sermons had been a crucial part of our lives in the women's section. I had felt sure I would be greeted and embraced as an old friend by him and his daughters. Instead, I had encountered a strange suspicion and hostility that vanished only when we could recall fragments of our shared past—Gladys's delicious sandwiches, the wonderful bond we had all felt at the Shield of Young David.

As I thought of the divide between me and the Ruben girls, the chasm between my world as a professional woman in New York and theirs as mothers and grandmothers tending to large extended families in this secluded Hasidic enclave in Jerusalem, I realized that theirs was the life I could have led—that would have been mine had I listened to Mrs. Menachem. And as I pondered the absence of children and grandchildren, along with the loss of community and all the sustenance it can provide, I felt shaken and no longer sure I had picked well and not at all confident in my chosen path.

*S*till trembling from my encounter with my old rabbi, I wasn't sure what to expect as I made my way to the West Bank settlement of Kiryat Sefer—the Town of the Book—where I had come to meet Celia's younger brother, Moshe.

As I stepped out of my taxi, I was met by a tall man in rabbinical garb. Moshe Garzon had an amiable smile and while he couldn't hug me—not in his culture—he managed to be embracing in his manner and greeted me enthusiastically. As Moshe took me inside the home he shared with his wife and family, I felt none of the wariness I'd detected in Rabbi Ruben's household. Occasionally, I'd catch glimpses of some of Moshe's eleven children scampering about, obviously curious about me.

There was even a Celia look-alike, a pretty little girl with dark hair and a mischievous smile who kept running over to us.

Moshe seemed very excited as he spoke—he, too, was an author he told me proudly, pointing to shelves lined with his books. He divided his time between his scholarly research and the rabbinical school he helped run. His wife who joined us was a serious, soft-spoken woman; she worked with Orthodox victims of spousal abuse.

Without any prompting from me, Moshe began to talk about our childhood synagogue and one of the central incidents of my life.

I was "the girl who made the rebellion" at the Shield of Young David, he said, conjuring up those long-ago Saturday mornings when my siege began. "You made a plan, you were going to sit in the men's section," he said. "I remember as if it were yesterday, because I was thinking, wow, this is unbelievable, those girls."

He had watched on the sidelines as Sabbath after Sabbath we'd move our chairs a bit more aggressively inside the men's sanctuary. He'd wondered if we were actually going to get away with it—he even recalled thinking there was a chance we would pull off the impossible. But then came the morning of the hue and cry, when we'd penetrated too far into the sanctuary, and the men had yelled, "That's it, girls, go back." He had seen us as we marched despondently back into the women's section.

Moshe spoke lucidly, thoughtfully, analytically, like the great rabbi I realized he had become. He wasn't in the least bit judgmental—he didn't condemn me or suggest I had transgressed any laws. Rather, he simply offered his own assessment of what went wrong.

"I will tell you why your plan didn't work," he said, growing excited, as if on the verge of solving an old Talmudic riddle. "You went too fast." Perhaps if I hadn't grown overly confident, he suggested, if I hadn't insisted on going too deeply inside the sanctuary, my scheme could have worked. The men at the Shield of Young David could potentially have made their peace with a group of little girls sitting in their midst. I listened to him, amazed: I had no illusions that Moshe believed in a synagogue without dividers, and I was sure that if I pressed him, he could have explained to me the intricacies of the laws governing the separation of men and women. But he also struck me as profoundly sweet, almost childlike, willing to go along with my girlhood fantasy. I also felt that he was trying to console me somehow for that long-ago debacle, perhaps rid me of the trauma I still felt.

I had only one more question: What had he thought all those years ago when he was watching, watching?

"I thought that you were a very courageous little girl," he replied, and he was smiling.

I said good-bye to Moshe and embraced his wife and children, including the adorable Celia look-alike. As I walked out into the balmy night air of the Town of the Book, I realized that I was feeling both elated and at peace. I had found a measure of solace at last at the hands of this gentle rabbi and childhood friend. It was as if he had sung to me the Verse of Consolation, that passage we recite when the Fast of Lamentation is over—*Rejoice, Rejoice and be comforted*—and we have ceased our mourning over the destruction of Jerusalem and have begun to rebuild our broken city.

A s a little girl, the divider had been the most visible, jarring aspect of a way of life filled with hundreds of strictures and ordinances—not simply where to sit and where not to sit inside a house of worship, but also what to eat, what to wear, when to work and when to stop all work.

There was a simple premise to my life as it unfolded behind the magical little barrier that seemed to shield me from all harm. As long as I sat in that women's section—as long as I didn't wander outside of it—I remained miraculously safe.

It never occurred to me to wonder if the men might be looking our way and wishing they could be with us in our haven.

The divider was long gone and with it, that feeling of complete serenity, the deep, abiding conviction that all would be well. Since its collapse I had come to see the world as a place of extraordinary, breathtaking peril. Without a divider to protect me, I was plagued by a perpetual feeling of danger—of fearfulness and timidity—all qualities that were foreign to me during my years at the Shield of Young David.

I had once felt supremely invulnerable. When I had stormed the men's section, I had been certain that I could change an ordained system in place for generations, even centuries. I saw myself as the avenger of the weak and downtrodden prisoners of the wooden enclosure.

To my shock and bewilderment, I found the world beyond the divider deeply wanting in comparison to the world I had left behind.

What I'd failed to realize was that for the women of my childhood, the world within our closed-off area was every bit as rich and vivid as the universe beyond it; and the barrier in fact fostered and intensified feelings of kinship and intimacy. Inside was a world that was remarkably collegial and embracing and kind.

I would find myself forever yearning for that sense of absolute protection, that feeling of being watched over and loved that I had experienced, I realized, only once—only in those years when I sat with Mom at the Shield of Young David in its women's section.

Inside the
Pasha's Library

Head straight to Sakakini Palace and then cross over to Ibn Khaldoum Street. It will lead you to Sakakini Street. But BE CAREFUL—there are not one but two Sakakini Streets. You want the Sakakini that has the tramway. Your mother's alleyway should be on the left."

Sarah Naggar, who as a little girl loved to watch my mom walking down the street every morning on her way to work, had been meticulous and painfully precise in the way that only the old can be: She'd repeated the directions to me over and over again, and always with the same warning.

I had to get to the street with the tram. If I wasn't careful, I would end up in the wrong Sakakini Street, the wrong alleyway.

Now that Edith was gone, I yearned for any trace of her. By returning to Cairo, I hoped to discover some remnant of the life she had loved. I had been back in search of my father's lost city, traveling to Egypt after an absence of more than forty years. Now, I hoped to find elements of the very different Cairo Mom had known, far removed from the cabarets and casinos my father had relished in his sharkskin suits, a world of humble alleyways and quiet libraries and schools. I was also curious as to what had become of the pasha's wife, the woman who had beguiled Mom's youth and haunted her old age. I intended to visit L'École Cattaui, where Mom had taught and the pasha's wife had presided. *Who knows, maybe I will see what is left of the library my mother built,* I thought hopefully. I planned to stop by Le Sebil. And I was going to hunt for the Cattaui Pashas: Surely, this grand Cairene family had left its imprint.

Most important of all, with the help of Sarah Naggar, now nearly eighty, I was going back to my mother's house.

I walked up and down Sakakini, searching for L'École Cattaui. But there wasn't so much as a trace of the school left—not even the original building. When I asked neighbors about the school, they looked at me blankly. Several blocks away, in the Abbassiyah section, I came across the tall imposing structure that had once housed the Sebil, the Jewish communal school where my mother had begun her teaching career and Alice Cattaui had supervised meals and handed out pairs of *sandalettes* to needy children.

I climbed up the steep flight of stairs, pushed open the door, and was greeted by men in long robes. They were imams—Muslim preachers—and they were now in charge of the building. They peered at me curiously. Yes, they said vaguely, they'd heard that years back, this had been a school for Jews, but it was a school for Muslims now.

Disconsolate, I circled back to Sakakini and the palace. From afar, it looked as grand—and rococo—as ever, but when I made my way to its front gate, I realized it was as faded and decayed

as the rest of Cairo. No one lived there anymore. The magnificent
palazzo was abandoned. I'd heard there had been plans to turn it
into a museum, but like so much in Egypt, that dream was never
realized.

When I crossed back to Sakakini Street, I remembered Sarah
Naggar's admonition: I stopped and asked an amiable storekeeper,
a man who sold ballpoint pens—only pens—in a shop the size of
a broom closet, if this was indeed the street of the tram. I couldn't
see any trams. He smiled broadly and nodded. He had grown up
in the neighborhood and fondly remembered the tram. Cairo was
more modern now, he said sadly.

Only Mom's little alleyway had stayed the same. Dusty, forgot-

*Loulou on a visit to Haret el-Helwa,
the Alley of the Pretty One, where
Edith had lived, Cairo, 2010.*

ten, the Alley of the Pretty One—Haret el-Helwa—still had the low-lying houses dating back to the early twentieth century, with balconies that looked unchanged from the years when Edith and Alexandra would sit back and savor their *café turc* while chatting with neighbors and passersby.

I lingered, transfixed, in front of my mother's girlhood dwelling, suddenly filled with the hope that at any moment, she and my grandmother would emerge, that I'd see them strolling arm in arm on their way toward Sakakini Palace for their evening walk.

"Blessed be the Lord who revives the dead," devout Jews pray every morning and every afternoon and every night.

Here in Cairo it was possible to take the prayer literally.

From Sakakini I made the pilgrimage to Garden City, as Mom had done that day Alice Cattaui had extended the invitation of a lifetime: to come see the pasha's library and browse through its wondrous collection of books. Although the taxi ride took only minutes, it was still like crossing into a different world, a quiet rarefied universe distinct from the grimy bustle of Sakakini.

Garden City had retained elements of its former splendor. Although many villas had been torn down and replaced by ugly concrete modern buildings, there were still streets lined with the dreamy palatial mansions of old. I peered one by one at the ancient villas and crumbling palaces. I consulted local scholars and historians and guides to help me: Where was Villa Cattaui? What happened to 8 Ibrahim Pasha Street?

The villa had vanished. A tall building stood in its stead.

I also tried to inquire about the Cattaui family cemetery, where Indji was buried near the pasha and his wife. It, too, didn't exist anymore, I was told. The explanations were vague and unsatisfactory. It had been torn down. It had been replaced by a new urban project. It had simply disappeared.

It was as if every trace of this great family that had once dominated the life of the city had been expunged. There was nothing left of the Cattauis in Cairo.

Or next to nothing. While wandering around the cool marble sanctuary of the Gates of Heaven, the synagogue designed and funded by the Cattauis where my parents were married, I noticed a plaque or two bearing their name, along with lavishly sentimental tributes to the family.

And that was it; all I could find of the storied Cattaui Pashas were a couple of square slabs of pale gray marble, like headstones, inscribed with the Cattaui name as well as some sentences in gold letters extolling their extraordinary love and devotion to the Jewish community.

The Cattauis, it turned out, were also victims. As with my own family, as with tens of thousands of Jews of every economic and social strata, the ferocious revolution hadn't viewed them as true Egyptians either. Stripped of their standing, their homes, their titles, and many of their holdings, they hung on at first, then painfully made their way out of Egypt, settling in Paris and London, Geneva and Lausanne. Some married into European nobility while others found spouses in the haute bourgeoisie.

They had means, of course—far more than my family ever would. They were able to lead comfortable lives and befriend other expatriates.

And yet they, too, mourned their lost life. The grace and elegance of their new homes wasn't enough to console them for their loss. Perhaps as a way to adapt, to be a part of a new culture that was so much less tolerant than Cairo had once been, one by one the Cattaui children and grandchildren relinquished their Judaism. Once the leaders and protectors of all the Jews of Egypt, the Cattauis no longer considered themselves Jewish. It was as if the keepers of the flame had chosen to extinguish their own inner flame.

Of course, once upon a time in old Cairo, it was possible to be Jewish and a pasha. You could be Jewish and an aristocrat, Jewish and a friend to ministers and kings. But the world beyond was far less embracing—and far more contemptuous—of Jews than this ancient, devout Muslim city.

For the Cattauis to take their place among European nobility, it seemed prudent simply to surrender their Judaism. Some flourished in France in part *because* they forfeited their identity. Georges Cattaui, a nephew of the pasha, was a French diplomat and distinguished literary critic who gained renown for his studies of Marcel Proust. Georges fashioned himself as a devout Catholic and, at the end, quietly requested a Christian burial.

But there was one Cattaui who didn't leave Egypt and didn't abandon Judaism and remained faithful to both to the end. The pasha's wife never left Garden City. She stayed even after the Cattaui name had lost its luster and the new regime had decreed it had no use for pashas or their wives.

In her final years, Madame Cattaui, so stoic and independent throughout her life, needed help to get by. Nearly blind in her old age, she found it hard to function in that enormous house, although the cook and several of the maids and even her faithful chauffeur remained by her side.

Odette Harari, a Jewish housewife who had grown up across the street from Villa Cattaui, was hired as her companion. Every day, Odette would arrive at the mansion to meet with Madame Cattaui. Typically, the chauffeur would drive the two women to the Gezirah Sporting Club, a magical enclave of gardens and flowers and tennis courts located in Zamalek, an island in the middle of the Nile. In the blissful seclusion of the Sporting Club, where once upon a time high society had reigned and the pasha's wife had reigned over high society, the two women would sit and enjoy a cup of tea and some quiet conversation.

Back at the Villa, Madame Cattaui would have Odette read the papers out loud to her: She was especially fond of the society column that chronicled births, deaths, weddings, and engagements. She always listened carefully to the obituaries, and when a friend from her old circles died, she would dictate a graceful condolence note to their family. She used the dead person's honorifics, referring to them as Pasha or Bey or Effendi—no matter that the revolutionary government had banned all royal monikers.

Once in a while, as Madame Cattaui held on to Odette's arm, the chauffeur drove to the small family cemetery on the other side of Cairo. Even in her eighties, the pasha's wife had continued her pilgrimages to Indji's grave and she was still mourning the daughter she had lost half a century earlier. Alice Cattaui was so frail by then, barely able to walk from the automobile to the gravesite.

As her strength failed her, the pasha's wife didn't leave the villa, and the beautiful sofas and armchairs where duchesses and queens had once sat and paid homage were covered by ghostly dustcovers: There were rarely any visitors anymore.

Only her granddaughter Nimet stopped by on occasion. Ever the contrarian, Nimet had returned from abroad and settled in Egypt even as Jews were leaving in droves. Her marriage to another Egyptian Jew, the son of a pasha, had ended badly. She returned to Cairo, married a Copt, and tried to keep an eye on the woman who had always counseled her, "Never give in to despair."

In the end, the pasha's wife was confined to her bed. One afternoon in 1955, Odette arrived to find Madame Cattaui was dying. Her granddaughter was at her side and understandably distraught.

Odette summoned her husband, but by the time he arrived, Alice Cattaui Pasha was dead. Mr. Harari began reading the Hebrew prayers of mourning, the Psalms and supplications the

pasha's wife had known only too well, since she had been reciting them from that day in 1908 when she had lost the child of her
heart, Indji.

At last I found Edith and the pasha's wife, though it took
another journey to another city, another continent, to
commune with a branch of the Cattauis. In the spring of 1969, Stephane Cattaui, the pasha's grandson, married a young Harvard-
educated woman named Maria Livanos. The lavish wedding in
New York recalled the Cattaui opulence of old. Guests flew in
from Europe, and the bride wore a tiara from which flowed a
twenty-foot train. The tiara sat on Maria's head, affixed with a
turquoise pharaonic-era bracelet first worn by the pasha's wife
in Egypt in 1922 at the opening of King Tut's tomb. The bracelet
had been a gift of Fouad, the sultan and ruler of Egypt who was
about to become king.

Maria came from a family with its own fabled past—a distinguished Greek shipping dynasty—but as she listened to her
husband's lyrical stories of Cairo, she realized that she had to
find a way to preserve the memory of the Cattauis. She began
to systematically organize the thousands of photographs, documents, letters, newspaper clippings, and mementos the Cattauis
had taken hurriedly out of Egypt and stashed pell-mell in cartons
and boxes. She put together photo albums and sorted through
the treasures retrieved from Villa Cattaui—sets of old china, silverware, gold-rimmed glassware bearing the distinctive Cattaui
initials, even odd pieces of jewelry, like the diamond-and-ruby-
studded pin Alice Cattaui Pasha had worn as lady-in-waiting to
Queen Nazli.

After Stephane became ill and died, the task of preserving the
Cattaui legacy was left entirely to Maria.

Yes, she told me, when we first spoke by telephone—she knew all about the pasha's wife and Indji, about the pasha and his library.

Would I like to come see for myself?

The invitation was irresistible: At last, I would find the Cattauis. I was going to learn about the library and the key and all that Edith had held dear. Within weeks, I had arranged to fly to Switzerland and to Maria Cattaui's residence.

The handsome chalet nestled in a village overlooking Lake Geneva seemed an improbable place for a memorial to a noble Egyptian-Jewish family, but that was in effect what Maria had created: a shrine in honor of the pasha and his wife.

Here were the delicate monogrammed dishes and gold-rimmed plates and cups and saucers and glasses and sets of silverware that Alice had used to serve the visiting dignitaries who had gathered week after week at Villa Cattaui. There in the foyer was the bust of old Yussef Cattaui Pasha. On the wall were illustrations of the various Cattaui palaces, one more shimmering and dreamlike than the next, and all bearing wonderfully exotic addresses—Shoubra, Khas al Dubara—addresses that conjured up a life of pleasure and absolute luxury, of all-night soirees with members of the royal family. In the albums were newspaper clippings along with photos going back to the nineteenth century—of young Cattaui and of Alice before she was even married, and then looking so formidable as lady-in-waiting standing next to the pasha, or attending functions with Queen Nazli.

And there were the dozens of portraits of Indji—Indji as a child, Indji as a pampered little girl, Indji as a young woman in white holding a rose. Even out of Egypt, the Cattauis had remained obsessed with her and had sought to preserve her memory.

As I wandered through the house, browsing at the statuettes and objets d'art taken from Cairo, I suddenly noticed the lock—the massive wrought-iron lock. It was embedded in a large glass door that led to the living room. I thought: how strange that the Cattauis had chosen to take a lock with them out of Egypt, along with their most precious possessions.

I imagined the Cattauis in their final days in Egypt, packing whatever seemed important and necessary. But then, at the last minute, someone had decided this lock was also indispensable, and they'd taken the trouble of having it removed from the heavy wooden door where it was ensconced, unable to stand the thought of leaving it behind.

Seeing it made me tremble: Was this the lock to the pasha's library? Where was the key?

There had indeed been a key once upon a time, my hostess informed me. Maria vividly remembered an ornate fantastical key that had accompanied the lock from Cairo. She had seen it and held it in her hand. But it was lost; it had been missing for years.

And then, it was as if the Messiah had come. Surely, he must have been there, in that corner of Maria Cattaui's chalet, when I spotted it—the library, the pasha's library. Mom's seemingly mythical haven was real and situated in this house overlooking the Mont Blanc. There were all the books—the books that she had loved, the books that she had borrowed, the books that she may have helped acquire, the books that were the apogee of her arrogant years, because after them nothing much had mattered.

The contents of the original Bibliothèque Cattaui, many of the sixty thousand volumes Yussef Cattaui had collected over his life, had largely been sold or auctioned off, and yet—miraculously—hundreds upon hundreds, even thousands, of his collection of first edition, leather-bound volumes had been saved, rescued from the library in Cairo, transported to Switzerland, and brought here to be lovingly arranged and displayed by Maria on shelves that seemed to reach from the earth to the sky.

Here were the collected works of the Goncourt brothers and Flaubert, of Zola and Guy de Maupassant and Proust's *Swann's Way*. There was the collected poetry of Gerard de Nerval and the plays of Molière. The *Temptation of Saint Anthony*. *Germinie Lacerteux*. Victor Hugo's *L'Art D'Être Grandpère—The Art of Being a Grandfather*. *The Hunchback of Notre-Dame*.

As I fingered each volume, I felt a need to remove them from the shelf and hold them. I kept asking myself: Was this a book Edith had purchased for the pasha's library? Was this a novel she had selected? Had she leafed through this collection? Had she enjoyed that anthology? Brushing over the soft leather jackets, I felt as if I were touching my mother's hand, exactly as I had as a little girl, when she would not let me go, she would not let me go.

· ACKNOWLEDGMENTS ·

As readers of my first memoir *The Man in the White Sharkskin Suit* learned, illness has shadowed me since I was a child, and early on I came to appreciate what the French call the *bon docteur*, that marvelous creature who knows how to diagnose and treat a malady and uses instinct as much as scientific knowledge in the way that he or she approaches a patient. The idea for *The Arrogant Years* came in part from my encounters with such physicians. In recent years, both my husband and I faced the threat of major illness, and in my case doctors constantly talked about *sequelae*. Because of my bout with Hodgkin's at age sixteen, I was told that I was still paying the price—and could expect to keep on paying. That is how I became interested in the notion of *sequelae*—the aftereffects of a major malady that are more subtle, that aren't even visible and yet are still devastating.

I remain, above all, grateful to the physicians and institutions who have watched over us. The inimitable Dr. Larry Norton, a deputy physician-in-chief at Memorial Sloan-Kettering, was anxious that I work on this book, even as illness struck again and I had lost both the will and faith to do very much. Each time that I went to see Dr. Norton for a checkup, he would ask about my progress with *The Arrogant Years*, and he gave me the confidence I needed to apply myself to it. Dr. Monica Morrow briskly, efficiently, and with breathtaking competence took over my case so that I *could* work. Finally, there was John Gunn, Memorial's chief operating officer, who has helped oversee an extraordinary institution that has been, for better and for worse, a part of me since I was sixteen. I am profoundly grateful to him for the access he offered. And, of course, Dr. Burton Lee remained a friend and medical counselor throughout my recent ordeals, exactly as he's been since I came to know him when I was a terrified teenager faced with a shattering diagnosis in the spring of 1973. Dr. Lee has been the ultimate *bon docteur,* and while no longer with Memorial he is to my mind a grand and indelible piece of it.

I am so fond of all of them and the miracles they perform day and night.

I also wish to pay tribute to my husband's doctors. The book couldn't have been completed without my husband or his doctors. Dr. Michael Grasso, formerly of Saint Vincent's—a truly blessed place that no longer exists—was a force of nature in attending to Feiden. Dr. Grasso applied his wondrous energy and talent to my husband's care during a time of unspeakable darkness. I have tremendous love and regard for him. Dr. Robert Motzer, of Sloan-Kettering, graciously agreed to become his oncologist, and each appointment showed me what a careful, thoughtful *bon docteur* he is. Both are remarkable men and I am not sure what we would have done without them. My internist, Dr. Jerome Breslaw, has watched over me so carefully these many years, and I am indebted to him and his entire practice, including his wife and office manager, Maddy, and their assistant, Maryann Cunningham. Dr. Ronald Schwartz helped me to make this book my priority.

Two people were central to the writing and editing process: my agent, Tracy Brown, and my editor, the sublime Lee Boudreaux. Tracy was the first to recognize the importance of the story of the lost Jews of Egypt. With his guidance I embarked on one memoir, then another, telling this story of exile and loss through the framework of one troubled family—my own. I have always trusted his literary judgment and have found him to be a superb sounding board when it comes to figuring out what makes a book work.

Lee Boudreaux devoted herself tirelessly, hopelessly, and completely to editing this manuscript. I felt at times as if she understood me better than anyone I'd ever known—which I suppose is what makes her such an amazing editor. I had a sense that I was in the hands of someone who didn't simply grasp what I was trying to say but also knew what I had *failed* to say, and I found her insights and her edits of the manuscript illuminating and ultimately transformative. Lee's enthusiasm is surely one of her great gifts.

Neither *The Arrogant Years* nor *Sharkskin* would have been possible without Ecco and Daniel Halpern. Publisher, poet, and lover of the Levant, Dan has turned Ecco into a literary treasure-house. Long before the world became riveted by Egypt and events in the Middle East, Dan and his staff had recognized the region's importance as a source and setting for literature. I must also give special credit to Abigail Holstein, the gifted associate editor who helped to shepherd both of my books; Abby is a wonder—creative as well as efficient, with a magical ability to pull a manuscript together.

Ghena Glijanski, formerly of Ecco, was both editor and psychologist extraordinaire. I am certain that I could not have written this book without her kindness, talent, encouragement, gentle counsel, sweetness, and editorial spark.

I depended on sources in many countries to research *The Arrogant Years*—the story that I wished to tell was in the hearts and memories of people who had been dispersed to the far corners of the Earth.

Switzerland: It was in a serene mountain village overlooking Mont Blanc that I found the magical Maria Cattaui, who proved to be so essential in helping me locate the keys to the pasha's library—

and to my mother. Maria was the kindest hostess imaginable. She welcomed me into her house and gave me access to its treasures, notably the books and photo albums of the Cattaui clan, which she had lovingly collected and preserved. She is simply a magnificent person and I was fortunate to meet her through a dear mutual friend, the late Congressman Steve Solarz. Solarz was an extraordinary supporter and well-wisher, and simply a wonderful human being, and many of us miss him terribly.

Yussef "Sousou" Makar, an old friend of the Cattaui family, was so generous with his time and ideas. He was also one of the most loving people I have ever known. He exhibited such passion, whether he was showing me the frayed, old patient roster of his late father, a top surgeon at L'Hôpital Israélite, or escorting me and my husband to dinner in Cairo's venerable Automobile Club. He also offered fascinating insights into the community of expatriates that found its way to Geneva and Lausanne—people who had left Egypt with their fortunes, yet felt every bit as lost as my family in the world beyond Cairo.

It was also my honor and privilege to meet Ahmed Fouad II, the son of the late King Farouk. I found him to be a gentle and thoughtful person, so sad about what had happened to his beloved country. As he spoke, I kept thinking how much my father had loved his father, and what a magical place Farouk's Egypt had been.

Israel: Thousands of Egyptian Jews found a home in Israel after they were forced to leave Egypt, and it was there that I located individuals able to help me reconstruct much of my mother's early life. Chief among them was Sarah Naggar Halawani, who was able to direct me to my mother's alleyway and helped me recapture the world of Sakakini and the Alley of the Pretty One, down to how the coffee tasted in the afternoon on the balcony and how it felt when that marvelous breeze blew in from the Sahara every evening. Sarah was so generous with her memories, and so vivid in her descriptions, that I found myself brought back to Edith's world, able to picture Mom as a beautiful young woman, proudly walking to her job as a teacher at L'École Cattaui.

Avi Nesher, the Israeli (and American) film director was a wonderful friend, encouraging me to get cracking on *The Arrogant Years*. I loved his passion and enthusiasm and when I was faltering, he urged me to focus on finishing. I value the friendship I have enjoyed with him and his wife, Iris; he is such a towering talent, such an extraordinary filmmaker, and I feel honored to be in his orbit.

My sister's childhood friends, Maggy, Etty, and Eva Wahba, were invaluable to my research. Maggy Wahba was so kind to share with me her impressions of my mom. Suzette had always adored the Wahba sisters and saw them as her own family. And I came to love them, too; they gave me a window on this vanished universe we had once shared.

I found Nimet Cattaui, granddaughter to the pasha and his wife, residing in a nursing home in Ashdod. She dressed with extreme simplicity and shared a small room with another woman, but I did notice one sign of her family's former greatness—over her small cot she had hung the photograph of Yussef and Alice Cattaui Pasha on the occasion of their fiftieth wedding anniversary in Cairo. Nimet was incredibly dignified despite her difficult circumstances. How far she had come from Villa Cattaui. . . . Yet she managed to conduct herself with such grace. She was exceedingly kind to share with me her memories of the pasha and his wife, and that lost glittering period in old Cairo.

Rabbi Moshe Garzon, my friend Celia's younger brother, proved extraordinary in helping me pull this book together. He was so kind and hospitable, as was his wife and family, and I have enormous affection for them.

France: I was able to interview several members of the Cattaui family in Paris. I am very grateful to Michel Alexane, Nimet's son and the pasha's great-grandson. It was Michel who led me to his mother in Israel and offered me insights into this magical and storied family. In charming encounters at the Café de Flor, Indjy Cattaui-Dumont, the pasha's magisterial granddaughter who was named after the tragic and mythical Indji, offered me many colorful stories about her upbringing in Villa Cattaui.

One of the great surprises of the last couple of years was hearing from members of the Dana clan. My grandmother Alexandra's family, the Danas of Alexandria and Cairo, had been largely unknown to me. Rachel Dana, whose father, Edgar, was Alexandra's brother, proved to be a remarkable resource. She had vivid memories of my grandmother and how excited she was when Edith became engaged to a wealthy, handsome older man, my father, Leon.

I had a wonderful encounter with Odette Harari, who had worked as the companion to the pasha's wife in her later years. Mrs. Harari provided extraordinary insights into the character of this grand lady who had so haunted my mother. I found Mrs. Harari's memories of Madame Cattaui blind and alone in Villa Cattaui simply haunting. I am so grateful to her son David Harari for helping to arrange a meeting between me and his mom. They helped so much with this book.

Milan: I have depended on the memories of my remarkable cousin Salomone Silvera for both memoirs. Salomone had recreated the world of my father on Malaka Nazli Street for *Sharkskin,* and he gladly shared with me the glimpses he had of my mother arriving at the house as a timid and frightened young bride, who was clearly not used to my dad's ways. My memoirs have brought me closer to Salomone and his wife, Sally, who sadly passed away as I worked on this book. I am also indebted to Davide Silvera, his youngest son, who was such a delight and who shares my intense interest in our family's quirky history.

Enrico Picciotto, Salomone's friend who also spent the War in Egypt, was able to share delightful stories of those years when the Nazis had taken over Italy, and he and his family had found a safe haven in Cairo. My most enjoyable research was done at fine Milanese restaurants in the company of Salomone and Enrico.

London: In the course of one magical evening at the Wolseley, Sir Ronald Cohen offered thoughtful insights into the Egypt of the Fouad era that helped shape my research into *The Arrogant Years*. I so enjoyed getting to know him and his wife, Sharon Harel, and delighted in their vivacity and intellectual spark. Sir Cohen, of course, is the Egyptian-Jewish expat extraordinaire, and it was touching to see how connected he remains to our shared heritage. I also benefited from the help of their friends Guy Naggar and his wife, Marion. Ronald's and Guy's thoughts about Fouad persuaded me that Egypt in the 1920s and 1930s had been an even more liberal and tolerant society than it was under King Farouk. Their sweetness and encouragement gave me the confidence to proceed with my undertaking.

Cairo: One of my most important accomplishments of recent years is that I have made friends, true friends, in Egypt, and in the process I have reclaimed, bit by bit, corners of my parents' beloved city. I would like to cite Hind Wassef, owner of the Diwan bookstore, perhaps my favorite destination in Cairo. I have never felt as heady or fulfilled as when I have spoken at Diwan, and I consider its owner one of Egypt's most gracious and promising citizens. I am also indebted to Samir Raafat, one of Cairo's most eloquent, thoughtful, and charming intellectuals. I cherish my bond with Samir: He is surely one of the most knowledgeable people on this vanished Egypt I love so much, and his research has simply been breathtaking.

I am grateful to America's former ambassadors to Egypt, Francis Ricciardone and Margaret Scobey. Mr. Ricciardone helped open the doors for me in Egypt. And I shall never forget the gracious evening that Ambassador Scobey arranged for me at the Embassy, where I met people who would prove invaluable as I pursued my research for *The Arrogant Years*.

Professor Yoram Meital, of Ben-Gurion University, was a wonderful resource in explaining the Fouad era to me. As we sat in the courtyard of the Marriott Zamalek sipping drinks and enjoying the pleasant surroundings, Professor Meital, an expert on Egyptian Jewry who was visiting from Israel, shared with me stories of Hoda Shaarawi and made me understand how pivotal the episode when she tore off her veil had been in transforming modern Egypt. I realized that for a brief, exhilarating period in the 1920s and 1930s Egypt had been a place of tolerance and progressive ideals for women and men both. These values were, of course, crushed by the Revolution of 1956 and decades of rule by the generals, and now, sadly, the veil has returned to Egypt.

Abdel-Hamed Osman, a young Egyptian with deep ties to the Jewish community, took me personally around Sakakini. Together, we found my mother's alleyway, surely one of the most moving experiences I have ever had. He was so gentle and understanding that I was reminded of what my family always missed the most about Egypt: the loving kindness, the intense humanity. Finally, I have been so touched by Carmen Weinstein, the indefatigable head of Cairo's dwindling Jewish community. Ms. Weinstein has continued to work so hard to preserve whatever she can of Jewish culture in Egypt. Her work getting Egypt to refurbish and, in effect, rebuild the venerable Maimonides Synagogue in the Jewish Quarter was a miracle worthy of Maimonides himself, as is her simple regular outreach to the remnants of our aging Jewish population.

America: George Rohr and the Jewish Book Council played a profound role in my ability to realize *The Arrogant Years*. In one blissful night in Jerusalem, George handed me that miraculous prize he had established in honor of his father, Sami, which enabled me to travel freely and work as I needed. I came to treasure his gentle counsel and encouragement, and he emerged as a wonderful friend. Then there was the miracle of the Jewish Book Council itself—the organization headed by Carolyn Hessel that now dominates the Jewish literary world in America. Carolyn has been my personal guardian angel. Filled with love and enthusiasm, she would not rest until I fulfilled the promise of the Rohr Prize and was well on my way to working on this companion volume. The Council—with Carol Kaufman, Naomi Firestone-Teeter, and Miri Pomerantz Dauber—functions like a family, and they became my family.

Through the Rohr Prize I have met many wonderful people who helped with this undertaking. Jonathan Sarna of Brandeis University was able to shed light on a period in my family's history that I had always found puzzling—the mystery of my ancestors, the Laniados of Aleppo, and their embrace of the false messiah Sabbetai Tzvi. Professor Sarna offered me a new way to look at my flawed relatives of old, persuading me to be more compassionate. He is a font of knowledge and one of the nicest people on this Earth. Yossi Klein Halevi of Israel and Brooklyn told me captivating stories about being a young member of the Jewish Defense League. It was in the JDL that he met someone from my old Syrian community, a young man named Joseph Hannon; I am so grateful to Yossi. Joseph Telushkin has been such a caring and amiable force among the judges as well.

I adore Nessa Rapoport's enthusiasm, sweetness, and terrific insights; she is a national treasure. I am also indebted to Daisy Maryles and Rela Mintz Geffen, also of the Rohr Prize judges panel, who approach their task with so much love.

I am honored to have come to know two other Rohr Prize

judges: Deborah Lipstadt of Emory University, whose work continues to shake us and surprise us, and the soulful Ruth Wisse of Harvard University. Finally, I cherish my bond with Sam Freedman and Ari Goldman, both of Columbia University.

My family helped me with much of my research. César, my oldest brother, conjured up key episodes in my family's past and was very generous in sharing with me his own memories and perceptions of our mother. His wife, Monica, culled together hundreds of family photographs taken over the years. My nieces, Caroline and Evelyn, were intensely supportive, each in their own intensely gracious way; they are very dear to me.

My sister, Suzette, was able to fill in details about my mother before I was born, including summers in Ras-el-Bar and in Alexandria; I appreciated her meticulous and detailed rendering of vacations at the Villa LaLouche. I am grateful to her and her son, Sasha, now a talented physician.

My book required me to take a journey deep into my own past, back to the women and men of the Shield of Young David, my childhood synagogue, whose name I translated as the Shield of Young David. I wish to express my regard for Marlene Ben Dayan, who once sat with me in the women's section of our little shul in Brooklyn. I benefited from Marlene's keen mind and terrific memory. In between cooking lessons, where she taught me to prepare some favorite dishes of my childhood, it was blissful to sit in her magical kitchen near Ocean Parkway and reminisce about our shared past. Marlene and her husband, Avi Ben Dayan, were always so hospitable, and I turned to them again and again in this difficult undertaking, appreciative of their insights into the world of Syrian Jewry caught, like me, between the traditions of old Aleppo and the temptations of modern New York.

I enjoyed being reunited with one of my favorite childhood friends, Celia Weinstein née Garzon. It was wonderful to share memories with her of her mother, tender Madame Marie, and I was reminded of why my mom had so loved the Garzon family and of how she had found in Celia's mother a haven from the brutish, new American culture that neither woman fully understood.

My cousins Rachel and Pico Hakim and their daughter, Rosette, were embracing and helped me as much as they could. They were so gracious and I love them dearly. Lily Halawani, my mom's niece, offered valuable insights into Edith as a young girl and gave me a vivid picture of my mother's bond with my grandmother Alexandra, conjuring the two walking arm in arm through the streets of Sakakini.

Desi Sakkal, founder of the website HSJE (Historical Society of Jews from Egypt) was wonderful as usual, answering any questions I had on the topic that, like me, he finds most interesting on this Earth—our lost Egyptian heritage.

Fortune Cohen—whom I had known as Fortuna Eddi as a young girl—was able to piece together for me the story of her late sister Gladys, whom I had loved so much in the women's section and found so moving. I was struck by what a sober, thoughtful, and purposeful woman Fortune had grown up to be, and I was deeply appreciative of the memories she proffered about her sister, one of my favorite people behind the divider.

When I began to research my mother's years at the Brooklyn Public Library, I discovered a group of people who reminded me of why Edith had loved the library so passionately and so much. I was moved to rediscover Rabbi Sam Horowitz and touched that *le gentil petit* Ed Kozdrajski, "Sweet little Ed Kozdrajski," and his wife, Maureen, both returned to their old stomping grounds to meet with me. Stephen Akey was especially kind to invite me back to the catalog department, and provided me with a detailed look at the work of a cataloger. Together, we mourned those who were no longer with us. Looking around those careworn desks, I wanted to cry out: Please let me stay here and work among you, in the place where I feel closest to Edith.

It was such a delight to find some of my old classmates from the Berkeley Institute, now the Berkeley Carroll School. Wendy Gold, who had been my closest friend when I was a student there, tried hard to conjure up memories of our ninth-grade class and what it was like to live through the changes that 1969 and 1970 wrought, not simply for us at Berkeley but for any young girl coming of age in that turbulent period. It was lovely to reconnect with her, and I found Wendy as thoughtful as when we'd shared a classroom so many years ago. Betsey Raze, she of the vivid mind, see-through blouses, and edgy views, struck me as more gentle than I'd remembered her—still filled with verve but softer somehow than that year when I eyed her with a mixture of admiration and envy. Betsy helped me piece together essential elements of that time. I enjoyed meeting up with Susan Weber Soros, whom I found ensconced in a magnificent town house and gallery on the West Side—home of the Bard Academic Center she had created as a center for the study of decorative arts. I couldn't help feeling wistful, realizing as we spoke that I had felt much more at one with those charming young girls of Berkeley than I had previously thought. I was able to contact my old headmistress, Mrs. Miller, and have some exchanges with her. She emerged as thoughtful and insightful. Finally, I much appreciated the efforts of Berkeley Carroll itself to dig out my records and help me piece together details from that lone year I spent within its gates.

I was also delighted to discover another Cattaui in America: Florence Sutter, the pasha's great-great-granddaughter, had moved to the Washington area and she showed up one evening at one of my readings. She struck me as such a thoughtful and sensitive young woman, and since that chance encounter, she and I have corresponded occasionally on this illustrious family of hers. She tried to help me as much as she could, as did her mom in Paris, and I appreciated their insights very much. Ed Adler, a Vassar alumnus, helped with some crucial memories of the period as well.

There were also several personal friends who helped in my being able to tackle *The Arrogant Years*. Janet Davis and Karen Merns accompanied me to Cairo on what proved to be a seminal trip for my research. Janet has been such a kind and sensitive friend, and I was so moved when she insisted on returning with me to my childhood home on Malaka Nazli Street. Karen's energy and enthusiasm were critical in getting me to undertake this final, crucial piece of research, and I am deeply grateful to her as well. Joann Lublin, my colleague at the *Wall Street Journal*, where she is a dazzling and indefatigable reporter, was there every step of the way, offering me the support and psychological courage I needed to persist and encouraging me when I was faltering. I am not sure what I would have done without Joann; she is a remarkable journalist and person.

Arthur Gelb, the éminence grise of the *New York Times*, its former cultural czar and managing editor, has been simultaneously father, friend, protector, adviser, and editor. It is not simply the help he provided on both books—he was extraordinary with *Sharkskin*—but his actions as a friend when I was ill. It was Arthur more than anyone who shook me out of my confusion and lethargy and persuaded me to do what I must to seek out the *bons docteurs* who proved so essential to my recovery. He is a national treasure and I owe him and Barbara, his wife, so much.

Jack and Isabel Biderman of Southampton consistently offered words of encouragement. They are both very dear to me.

As I worked on the chapters of my college years, it was wonderful to rediscover Laurie Wolf. Now married to Eli Bryk, a talented New York orthopedic surgeon, Laurie still had that spark and enthusiasm I had found so endearing when I'd first known her. How marvelous it has been to come together with her and Eli and their six amazing children, who with their sparkle and sense of fun always make me think of a family out of Tolstoy—one of his iconic happy families. Jacki Bryk, Laurie's youngest daughter, became my literary assistant and has worked energetically to help market both *Sharkskin* and now *The Arrogant Years* through social media.

In Sag Harbor, Canio Books is one of the most fascinating literary spots in the Hamptons. Its owners, Maryann Calendrille and Kathryn Szoka, are great women who have done so much for the cultural scene on the East End. Both have been tremendously supportive of me and my undertaking and, like their store, they are nothing short of wondrous. Romany Kramoris of the Romany Kramoris Gallery has been a wonderful, joyous friend, first of *Sharkskin* and now *The Arrogant Years*. Lynda Rosner at the Sag Harbor Variety Store on Main Street, was such a loving force, always offering words of support and reassurance.

Rabbi Rafe Konikov of Chabad of Southampton gave me a haven where I could pray and think and remember my parents. He and his wife, Chani, have created such a special place at 214 Hill Street. They are both exceptional people who have worked so intensely to create a Jewish community of faith and hope and religious observance in the Hamptons.

The Arrogant Years is filled with religious lore, and I turned to Rabbi Rafael Benchimol of Manhattan Sephardic Congregation with questions about the meaning of obscure fasts or other historical events. He always offered thoughtful insights—as when he told me the story of Rabbi Tsadok, the holy man of ancient Judea, whose passion for fasting I found of critical importance in shedding light on my mother. Whenever I found myself stumped by a religious question, I turned to Rabbi Benchimol and was always impressed by his sensitivity and knowledge. And in these years when my husband and I have been under the shadow of illness, he has been much more than a rabbi—he has been a caring friend and counselor, as have members of the synagogue, such as Leah Iny, Kim Amzallag, and the sublime Stella Issever. These women were the essence of tenderness and caring. Whenever my world seemed darkest in the last couple of years, I sought out Rabbi Benchimol and Manhattan Sephardic for sustenance and strength.

Robert Thompson, who took over as editor-in-chief of the *Wall Street Journal* when it was acquired by News Corp., was deeply gracious in blessing this undertaking and allowing me to go work on this book. I am utterly grateful. I am especially indebted to Alix Freedman of the *Journal* for granting me the crucial leave that I needed for *The Arrogant Years*. Alix, who has shattered many dividers herself over the years, has been an extraordinary supporter and friend. Joe White, my editor, has been a sensitive and important force throughout the process. Jim Pensiero and his assistant, the wonderful Theresa Pfeifer, helped with the intricate logistics of my leave and made it possible and manageable: I am so grateful to them. And in tech support, Sylvia Burgos and Willie Bennett were invaluable and rescued me time and again.

In the more than two years that I worked on *The Arrogant Years*, no one was more marvelously supportive than Grace Edwards, my teacher at Marymount and then at Hunter College. Her gentle critiques were crucial in shaping this manuscript. Lewis Frumkes, the head of Hunter's newly created Writing Center has been such an important force over the years. I am grateful to him and to Hunter's president, Jennifer Raab, for establishing a center with the potential to mold and encourage promising authors of different ages, backgrounds, and venues. My classmates Chuck Beardsley, Vicky Myers, John-Paul Nickerson, and Janet Davis offered insightful critiques, which were always filled with such tenderness and love.

Finally, my gratitude goes to Douglas Feiden, my husband, for his incredible dedication to me and to the realization of this book. Though often struggling with deadlines of his own, Feiden readily gave up his Sundays to hear me read a chapter, to which he offered his remarkably insightful remarks and suggested edits. This book benefited from his sensitivity and sensibility. I would have been lost without his guidance and fine-tuned ear listening to me read passage after passage. He is an extraordinary person.

· SELECTED BIBLIOGRAPHY ·

Books: Fiction and Nonfiction

Aldridge, James. *Cairo: Biography of a City*. Boston: Little, Brown and Company, 1969.

Benin, Joel. *The Dispersal of Egypt's Jewry*. Berkeley: University of California Press, 2004.

Cattaui, Georges, ed. *Marcel Proust: Documents Iconographiques*. Preface and notes by Georges Cattaui. Geneva: Pierre Cailler, 1956.

Cooper, Artemis. *Cairo in the War, 1939–1945*. London: Penguin Books, 1989.

Hughes, Pennethorne. *While Shepheard's Watched*. London: Chatto & Windus, 1949.

Lababidi, Lesley. *Cairo's Street Stories: Exploring the City's Statues, Squares, Bridges, Gardens, and Sidewalk Cafes*. Cairo: American University Press, 2008.

Lambert, Phyllis, ed. *Fortifications and the Synagogue: The Fortress of Babylon and the Ben Ezra Synagogue, Cairo*. Montreal: Canadian Centre for Architecture, 1994.

Mahfouz, Naguib. *Palace Walk*. Translated by William Maynard Hutchins and Olive E. Kenny. New York: Anchor Books, 1990.

Miller, Avigdor. *Rejoice Oh Youth: An Integrated Jewish Ideology*. Union City, NJ: Gross Brothers Printing, 1962.

Mostyn, Trevor. *Egypt's Belle Epoque: Cairo and the Age of the Hedonists*. London: Tauris Park Paperbacks, 2006.

Sacks, Oliver. *The Man Who Mistook His Wife for a Hat*. New York: Touchstone Books, 1970.

Salinger, J. D. *Raise High the Roof Beam Carpenters and Seymour: An Introduction*. Boston: Back Bay Books, 1955.

———. *Franny and Zooey*. Boston: Little, Brown and Company, 1955.

Schmitt, Eric-Emmanuel. *Monsieur Ibrahim and the Flowers of the Koran*. New York: Other Press, 2003.

Sinoue, Gilbert. *Le Colonel et l'Enfant Roi: Memoires d'Egypte.* Paris: Editions Jean-Claude Barres, 2006.

Stadiem, William. *Too Rich—The High Life and Tragic Death of King Farouk.* New York: Carroll & Graff Publishers, 1991.

Sutton, David. *Aleppo: City of Scholars.* New York: Mezorah Publications, 2005.

Sutton, Joseph. *Aleppo Chronicles: The Story of the Unique Sephardim of the Ancient Near East; In Their Own Words.* New York: Thayer, Jacoby, 1988.

Newspapers, Magazines, and Websites

Associated Press. "Shepheard's Hotel Is Reported Ruined—Americans, Escaping Through Mob, Tell of a Flaming Up, Possibly from Grenade." *New York Times,* January 27, 1952.

Daniel, Clifton. "Farouk Asserts his Kingship." *New York Times,* February 17, 1952.

Kaufman, Michael. "Two At Columbia Indicted in Sale of Cocaine." *New York Times,* December 2, 1973.

Oudiz, Albert. *"L'Ecole de Mon Enfance: Moise de Cattaoui Pacha, Souvenirs d'une Enfance Studieuse et Heureuse."* [The School of My Childhood: Moise Cattaui Pasha, Remembrances of a Studious and Happy Youth]. Reproduced on HSJE (Historical Society of Jews from Egypt).

Raafat, Samir. "Dynasty: The House of Yaccoub Cattaui." *Egyptian Mail,* April 2, 1994.

Sabato, Rabbi Haim. "Our Exodus from Egypt, 1957 Version." Hamodia Story Supplement, Passover 2007. *Hamodia: The Daily Newspaper for Torah Jewry.*

Archival Material

The Black Record: Nasser's Persecution of Egyptian Jewry. American Jewish Congress, 1957, New York.